SPECTROSCOPIC PROPERTIES OF ND:

YAG LASER AND ITS PERFORMANCE

MATERIALS SCIENCE AND TECHNOLOGIES

Additional books in this series can be found on Nova's website
under the Series tab.

Additional e-books in this series can be found on Nova's website
under the eBooks tab.

MATERIALS SCIENCE AND TECHNOLOGIES

SPECTROSCOPIC PROPERTIES OF ND:

YAG LASER AND ITS PERFORMANCE

SEYED EBRAHIM POURMAND

IRAJ SADEGH AMIRI

ABDOLKARIM AFROOZEH

VOLKER J SORGER

AND

XI LING

nova
science publishers
New York

NOTICE TO THE READER

Library of Congress Cataloging-in-Publication Data

ISBN: 978-1-53613-147-5

Published by Nova Science Publishers, Inc. † New York

Contents

PREFACE

The purpose of this book was to investigate the temperature and input energy dependency of Nd:YAG laser performance pumped by flashlamp. A commercial laser rod Nd:YAG laser crystal was utilized as a gain medium. The laser rod was placed parallel to a linear flashlamp filled by xenon gas at 450 Torr. The Nd:YAG crystal together with the flashlamp was flooded with a coolant comprising of a mixture with 60% ethylene glycol and 40% distilled water, which covers a range of temperature from -30°C to +60°C. Spectroscopic properties of the Nd:YAG rod under pulsed flashlamp pumping was investigated from the output fluorescence spectrum of the flashlamp radiation and the Nd:YAG rod. The linewidth of each fluorescence line was measured for an estimation of an effective emission cross section and saturation intensity. The influence of temperature and input energy on a fluorescence emission cross section of Nd3+:YAG crystal was studied. The cross-section was found to decrease as the temperature and the input energy was increased. The inter-stark emission showed a Lorentzian line shape indicating homogeneous broadening. This was attributed to the thermal broadening mechanism of the emission line. The spectral widths and shifts of the emission lines for the three and four level inter-Stark transitions within the respective intermanifold transitions of 4F3/2→4I9/2 and 4F3/2→4I11/2 were investigated over the range of 0 to 75 J. The emission lines for the 4F3/2→4I9/2 transitions shifted towards a longer wavelength and broadened, while the positions and linewidths for the 4F3/2→4I11/2 transitions remained unchanged with the increase of input energy. Finally, the temperature dependence of quasi-three-level laser transitions for long pulse Nd:YAG laser was also investigated. The laser performances at both 938.5 nm and 946.0 nm were also found to be inversely proportional to temperature, and the slope efficiency was unchanged with temperature. The reduction was due to the mechanism of phonon scattering as well as a broadening effect while the temperature increased.

Chapter 1

INTRODUCTION OF FLASHLAMP-PUMPED ND:YAG LASER

ABSTRACT

In this chapter xenon filled flashlamp was used as a pump source because it is the most efficient gas at converting electrical energy into optical energy and it is cheaper than other gasses. An optical resonator was aligned in the simplest form contained two parallel mirrors placed in between the gain medium which provides feedback of the light. In this research laser performance of Nd:YAG laser pumped by flashlamp was investigated. This study basically focused on both aspects, theoretical and experimental work based on Nd:YAG laser system. A commercial laser rod Nd:YAG laser crystal is utilized as a gain medium. Finally, in the present research simultaneous oscillation of dual-wavelength Nd:YAG laser at 938 and 946 nm pumped by flashlamp was introduced. Besides, laser performance of quasi-three-level transitions at 938 and 946 nm versus temperature and input energy was studied.

Keywords: flashlamp, optical resonator, Nd:YAG laser, waveguides

1.1. OVERVIEW OF FLASHLAMP PUMPED ND:YAG LASER

The word LASER stands for Light Amplification by Stimulated Emission of Radiation. The principle of the stimulated emission in addition to absorption and spontaneous emission was first introduced by Albert Einstein in 1916. He explained that in the presence of the field of excited photons, other atoms were stimulated to emit additional photons. The frequency of the emitted radiation was related to the difference in the atomic energy levels [1].

Generally speaking, a laser constructed from three principal parts, a pump source, a gain medium, and an optical resonator which includes two or more mirrors, as shown in Figure 1.

In this research xenon filled flashlamp was used as a pump source because it is the most efficient gas at converting electrical energy into optical energy and it is cheaper than other gasses [2, 3]. The flashlamp power supply was based on the series simmer mode triggering method and energized the flashlamp with high energy. A Nd:YAG crystal is used as laser medium which is a neodymium (Nd^{3+}) doped yttrium aluminum garnet (YAG). The Nd:YAG can produce more than 30 laser lines in the near-IR spectral region [3].

An optical resonator was aligned in the simplest form contained two parallel mirrors placed in between the gain medium which provides feedback of the light. The mirrors were coatings to determine their reflectivity properties.

For terrestrial applications, laser systems are mainly used in the temperature range from -60 to 60°C. Other optical elements in a typical laser resonator (e.g., mirrors, beam splitters, etc.) show no variation in optical properties over a wide range of temperature. However, ambient temperature and the heat generated by the gain medium of flashlamp pumping lead to thermal broadening and shift of laser lines which seriously affect on gain amplification, threshold power, frequency stability, and thermal tunability of the lasers and obstruct the lasing performance [4-7].

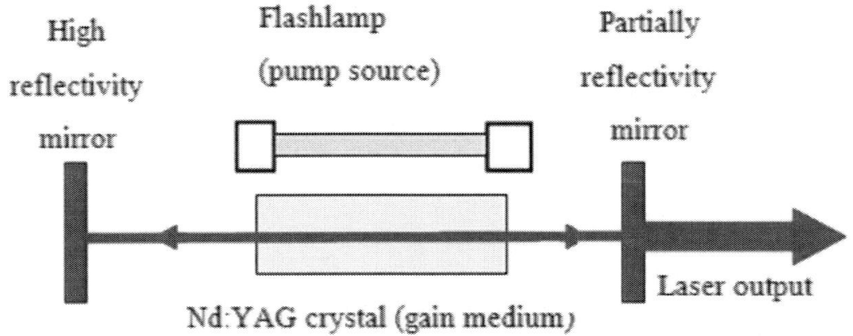

Figure 1.1. Schematic diagram of flashlamp pumped Nd:YAG laser.

In this research laser performance of Nd:YAG laser pumped by flashlamp was investigated.

1.2. INFLUENCE OF THE VOLTAGE ON A LASER DEVICE

Nowadays most simultaneous generation of multi-wavelength is generated by diode pump solid state lasers. The drawbacks of these operations are that they are continuous or quasi-continuous. These operations are difficult to control and in addition, they have low peak power [8].

Furthermore, in quasi-three-level lasers, in which the ground level has the significant thermal population, laser transition suffering from significant re-absorption. This phenomenon increases the internal loss in the gain medium and therefore population inversion would emerge at strong pump intensities. As a result, additional heat is generated by the crystal [9, 10].

Some parameters such as laser threshold, output power, internal loss, the linewidth of the laser lines are dependent on temperature, hence heat generation can dramatically influence the laser performance [11-13]. Furthermore, in most high power solid state lasers variation of emission cross section with temperature has a serious effect on the laser performance. Consequently, extra heat leads towards damaging optical components in intracavity and varies stability of the output energy over the temperature range of interest. Furthermore, in increasing the power of the laser normally involve in increasing the pump power. To date, no many works have been reported on the influence of temperature base on the voltage of flashlamp or the input energy of the laser. Therefore, the novel work claimed from this study is dependent on these results.

1.3. ANALYSIS OF FLASHLAMP AND ND:YAG ROD RESPECT TO DIFFERENT LASER PARAMETERS

Since Nd^{3+}: YAG laser performance is dependent on the temperature and pumping energy [14] careful investigation on the phenomena of linewidths and shifts of several lines of $^4F_{3/2} \rightarrow ^4I_{11/2}$ and $^4F_{3/2} \rightarrow ^4I_{9/2}$ intermanifold transition lines is required. In addition, changes in laser output energy with temperature also is a crucial issue in solid-state laser materials.

Accurate information on temperature and input energy effects would be possible to deal with the varying stability of the output energy over the temperature range of interest. Thus, for the investigation to achieve these objects, the following works were performed:

(i) Investigation of spectroscopy properties of flashlamp and the Nd:YAG rod in different temperatures and input energies.

(ii) Investigation on broadening effect and shifted off the center line at different temperature and input energies.

(iii) Estimation of stimulated emission cross section of the Nd:YAG quasi-three-level laser at different temperature and input energy.

(iv) Align of an optical resonator to produce a simultaneous generation of 938 and 946 nm wavelengths.

(v) Characterize the performance of quasi-three-level laser transitions for long pulse Nd:YAG laser at various temperature.

1.4. SCOPE OF THE THEORETICAL AND EXPERIMENTAL STUDY ON ND:YAG LASER SYSTEM

This study basically focused on both aspects, theoretical and experimental work based on Nd:YAG laser system. The system comprises of two parts. The first part is energizing a laser crystal and stabilizing the stimulating emission of fluorescence radiation. The second part is a measurement of fluorescence and laser radiation by the spectrometer.

A commercial laser rod Nd:YAG laser crystal is utilized as a gain medium. The doping level of the laser rod is 1 at. % with a dimension of 4 mm in diameter and 70 mm in length. The laser rod is enclosed in a ceramic reflector. The laser rod is placed parallel to a linear flashlamp filled xenon gas at 450 Torr. The rod is excited using side pumping technique with a homemade power supply [15]. The driver is triggered by simmer mode technique. A capacitor bank with the capacitance of 150 μF is charged by the maximum voltage of 1000 V thus the input energy is varied between 0 to 75 J.

Intensive emitted light by flashlamp dissipates heat to the laser rod and the flashlamp itself. Therefore, the Nd:YAG crystal together with the flashlamp is flooded with a coolant comprised of the mixture of 60% ethylene glycol and 40% distilled water. Such particular coolant covers the range of temperature from -30°C to +60°C. A thermocouple is connected to the heat sink of alumina ceramic to measure the temperature of the cooling system.

The fluorescence radiation after pumping is emitted at one end of the laser rod. The light is detected by a CCD camera. The spectroscopic properties were analyzed via a Wavestar version 1.05 software. The resolution of this detection system is 0.5 nm so it can resolve most of the transition lines appeared from the pumping rod.

The dual wavelength laser operation at 938 and 946 nm are generated simultaneously by using laser resonator mirrors precise coating at 946 nm with reflectivity at 75%.

1.5. Significance of the Experimental Evidence of the Emission Cross Section of Nd:YAG Crystal

A variety of laser materials has been developed, among which the most standard host is the yttrium aluminum garnet (YAG). Owing to remarkable laser properties of Nd^{3+} doped YAG crystal, such as high mechanical strength, thermal conductivity, optical transparency over a wide spectral region, adequate fluorescence lifetime for storage energy and high stimulated emission cross-section, it has been utilized for a long time in solid-state laser industry [16].

During the last decade wavelengths in the blue light region have attracted much attention because of their practical applications such as high-density data storage, color displays, Raman spectroscopy, underwater communication, high-resolution printing and medical diagnostics. One important way to produce blue lasers results from either second harmonic generation of 946.0 and 938.5 nm wavelengths or sum frequency generation of 946.0 with 938.5 nm in quasi-three-level Nd:YAG laser [17, 18]. Therefore, the stimulated emission cross section and output energy for two lines of 938.5 and 946 nm were quantified as a function of temperature and input energy. These observations are new and may contribute towards new design architecture of quasi-three-level laser systems.

Therefore, the stimulated emission cross section and output energy for two lines of 938.5 and 946 nm were quantified as a function of temperature and input energy. These observations are new and may contribute towards new design architecture of quasi-three-level laser systems. In addition, a critical parameter in such a laser system design is a fundamental understanding of the temperature and input energy dependent mechanisms or material properties that ultimately contribute to a change in laser performance.

Several works have been done on spectroscopic properties and stimulated emission cross section of Nd:YAG crystals at the major line of 1064 nm pumped by diode lasers. However, to the best of our knowledge, few studies have been conducted on the main lines of quasi-three-level lines at 938 and 946 nm wavelengths induced by a flashlamp pumped Nd:YAG laser. Therefore, in this book, we reported the experimental evidence that stimulated emission cross section of Nd:YAG crystal at these wavelengths was affected by the temperature and the input energy. In addition, input energy dependency of intensity, linewidth, and wavelength position of quasi-three-level and four level system transitions was investigated. However, to the best of our knowledge similar studies have not been reported up to now. Finally, in the present research simultaneous oscillation of dual-wavelength Nd:YAG laser at 938 and 946 nm pumped by flashlamp was introduced. Besides, laser performance of quasi-three-level transitions at 938 and 946 nm versus temperature and input energy was studied.

REFERENCES

[1] Natarajan, V., Balakrishnan, V. & Mukunda, N. (2005). "Einstein's Miraculous Year", *Resonance, 10*, 35-56.

[2] Perlman, D. E. (1966). "Characteristics and Operation of Xenon Filled Linear Flashlamps. Review of Scientific Instruments", *Review of Scientific Instruments, 37*, 340-343.

[3] Koechner, W. (2006). *"Solid-State Laser Engineering"*, 6th revised and updated version. USA: Springer.

[4] Rapaport, A., Zhao, S., Xiao, G., Howard, A. & Bass, M. (2002). "Temperature dependence of the 1.06-μm stimulated emission cross section of neodymium in YAG and in GSGG", *Applied Optics, 41*, 7052-7057.

[5]   Zhao, S., Rapaport, A., Dong, J., Chen, B., Deng, P. & Bass, M. (2005). "Temperature dependence of the 1.03 µm stimulated emission cross section of Cr:Yb:YAG crystal", *Optical Materials*, *27*, 1329–1332.

[6]   Sardar, D. & Yow, R. (1998). "Optical characterization of inter-Stark energy levels and effects of temperature on sharp emission lines of Nd3+ in CaZn2Y2Ge3O12", *Optical Materials*, *10*, 191–199.

[7]   Sardar, D. & Yow, R. (2000). "Stark components of 4F3/2, 4I9/2 and 4I11/2 manifold energy levels and effects of temperature on the laser transition of Nd+3 in YVO4", *Optical Materials*, *14*, 5-11.

[8]   Saiki, T., Nakatsuka, M., Fujioka, K., Motokoshi, S. & Imasaki, K. (2011). "Cross-relaxation and spectral broadening of gain for Nd/Cr:YAG ceramic lasers with white-light pump source under high-temperature operation. ", *Optics Communications*, *284*, 2980-2984.

[9]   Eichhorn, M. (2008). "Quasi-three-level solid-state lasers in the near and mid infrared based on trivalent rare earth ions", *Appl. Phys. B*, *93*, 269-316.

[10]   Lupeia, V., Aka, G. & Vivien, D. (2002). "Quasi-three-level 946 nm CW laser emission of Nd:YAG under direct pumping at 885nm into the emitting level", *Optics Communications*, *204*, 399–405.

[11]   Dong, J., Rapaport, A., Bass, M., Szipocs, F. & Ueda, K. (2005). "Temperature-dependent stimulated emission cross section and concentration quenching in highly doped Nd3+:YAG crystals", *Phys. Stat. Sol. (a)*, *202*, 2565-2573.

[12]   Turri, G., Jenssen, H. P., Cornacchia, F., Tonelli, M. & Bass, M. (2009). "Temperature-dependent stimulated emission cross section in Nd3+:YVO4 crystals", *J. Opt. Soc. Am. B.*, *26*, 2084-2088.

[13] J. a. D. Dong, P. (2003). "Temperature dependent emission cross-section and fluorescence lifetime of Cr, Yb:YAG crystals", *Journal of Physics and Chemistry of Solids*, *64*, 1163-1171.

[14] Pourmand, S. E., Bidin, N. & Bakhtiar, H. (2012). "Effects of temperature and input energy on a quasi-three-level emission cross section of Nd3+:YAG pumped by a flashlamp", *Chin. Phys. Lett.*, *21*, 094214.

[15] Zainal, R., Tamuri, A. R., Duad, Y. M. & Bidin, N. (2010). "Improvement in Ignition and simmer current supply into xenon flashlamp", *American Institute of Physics Procedings*, *1250*, 133-136.

[16] Kumar, G., Lu, J., Kaminskii, A., Ueda, K., Yagi. H., Yanagitani, T. & Unnikrishnan, N. (2004). "Spectroscopy and stimulated emission characteristics of Nd3+ in transparent YAG ceramics", *IEEE J. Quantum Electron.*, *40*, 231-235.

[17] Dimov, S., Peik, E. & Walter, H. (1991). "A flashlamp-pumped 946 nm Nd:YAG laser", *Applied Physics B: Lasers and Optics*, *53*, 6-10.

[18] Wang, C., Chow, Y. T., Yuan, D. R., Xu. D., Zhang, G. H., Liu. M., Lu. J., Shao, Z. & Jiang, M. H. (1999). "CW dual-wavelength Nd:YAG laser at 946 and 938.5 nm and intracavity nonlinear frequency conversion with a CMTC crystal", *Optics Communications*, *165*, 231-235.

Chapter 2

CONCEPTS AND PRINCIPLE WORKING ND:YAG LASER

ABSTRACT

We will discuss literature search on, oscillation 946 nm Nd:YAG laser. The first laser line at 946 nm in Nd:YAG active medium was produced by W. Wallace et al. The laser transitions at 946 nm and 938 nm has been noticeable because they revealed blue light region by nonlinear crystals. The first flashlamp pumped 946 nm Nd:YAG laser was reported by Dimov et al., in 1991. Continues wave dual-wavelength Nd:YAG Laser pumped by a flashlamp was reported in 1989. In this chapter properties and stimulated emission cross section of Nd:YAG crystals at room temperature have been studied extensively. This chapter aims to derive some useful equations for stimulated emission cross-section. Furthermore, main mechanisms theory of the spectral line thermal broadening and thermal shifts is investigated. Finally, Three, four and quasi-three-level laser systems were investigated.

Keywords: Nd:YAG active medium, oscillation 946 nm Nd:YAG laser, flashlamp, dual-wavelength Nd:YAG Laser

2.1. INTRODUCTION OF 946 NM AND MULTI-WAVELENGTH OSCILLATION ND:YAG LASER

This chapter will discuss literature search on, oscillation 946 nm Nd:YAG laser, simultaneous generation multi-wavelength oscillation, and effects of temperature on laser performance.

2.2. RESEARCH BACKGROUND OF FLASHLAMP PUMPED 946 NM ND:YAG LASER

2.2.1. A Short History of Oscillation 946 nm Nd:YAG Laser

The first laser line at 946 nm in Nd:YAG active medium was produced by W. Wallace et al. [1]. They used a resonator which comprised of a mirror and prism extremely reflecting at 946 nm and extremely transmitting at 1064 nm. Laser lines at 946 nm ($R_1{\rightarrow}Z_5$) and below it at 939 ($R_2{\rightarrow}Z_5$), 900 ($R_1{\rightarrow}Z_4$) and 891 nm ($R_1{\rightarrow}Z_3$) of Nd:YAG pumped by a pulsed xenon ion laser is obtained in 1972 [2]. These laser transition lines wind up in the stark sublevels of ground state $^4I_{9/2}$. A similar resonator as Wallace work also was used in this experiment.

The laser transitions at 946 nm and 938 nm has been noticeable because they revealed blue light region by nonlinear crystals. Fan and Byer [3] introduced the first quasi-three-level Nd:YAG laser. The laser was pumped by a diode laser and operated on the manifold $^4F_{3/2}$ manifold $^4I_{9/2}$ at 946 nm. Risk and Length [4] achieved 42-mW output power at 946 nm Nd:YAG laser line. They studied theoretically and experimentally on the quasi-three-level transitions. Also, blue laser light at 473-nm was achieved by Risk [5]. A nonlinear crystal ($LiIO_3$) was used as intracavity frequency-doubler. Afterward, various resonator designs and non-linear crystals have been exploited for frequency-doubling of 946 nm Nd:YAG laser [6-9].

The first flashlamp pumped 946 nm Nd:YAG laser was reported by Dimov et al., in 1991. The experiment was carried out using two linear xenon flashlamps at a temperature between 300-240 K. They had modified the cooling system and optical resonator design. The threshold energy obtained at input pumping of 75 J operating temperature of 298 K. The output energy of 3 mJ was recorded corresponding to input pumping of 90 J.

Barnes et al., (1997) achieved flashlamp pumped, room temperature Nd:YAG laser operating at 946 nm. They also studied on gain coefficient versus electrical input energy and performance of Nd:YAG laser at 946 nm at reflectivities of 99%, 97%, and 87% respectively.

In last decade many works have been carried out to obtain efficient laser diode pumped free running and Q-switching Nd:YAG laser transition at 946 nm [10-12].

2.2.2. Brief History of Simultaneous Generation Multi-Wavelength Oscillation

The first dual oscillation laser was generated by Bethea. He produced simultaneous wavelengths via Nd:YAG laser at 1064 and 1318 nm [13]. Then pulsed flashlamp laser in Nd: YLF laser [14] at both 1047 nm and 1313 nm, simultaneous multiple wavelengths lasing were reported. In 1987 dual wavelength via co-doped of Er^{3+} and Nd^{3+} [15] and co-doped of Ho^{3+} and Nd^{3+} [16] ions in YAG crystal has been produced respectively.

Continues wave dual-wavelength Nd:YAG Laser pumped by a flashlamp was reported in 1989 [17] and Wavelengths at 1064 and 1318 nm was obtained. Then Shen utilized Nd: YAP crystal pumped by flashlamp to obtained 1064 and 1318 nm [18]. He also compared simultaneous multiple wavelengths in different neodymium host mediums [19]. In 1999 a continues wave Nd:YAG laser operation at

938.5 and 946 nm pumped by a CW Ti: Sapphire laser was obtained [20].

Li et al. produced Nd:YAG laser line at 946 nm along with 1064 nm simultaneously for the first time [21] and overall output power of 2.5 W at the two transitions was presented. Hou et al. Subsequently, a Kr-flashlamp pumped Nd:YAG laser with simultaneous dual-wavelength oscillation at 1357 nm and 1444 nm was demonstrated in 2008 [22]. In addition a simultaneous dual-wavelength Q-switched Nd:YAG laser diode-side-pumped at 1318.8 and 1338 nm with the peak power of 43 kW was demonstrated in 2008 [23]. A Nd:YAG dual-wavelength laser pumped by diode laser oscillating at 1319 and 1338 nm was produced in 2009 [24]. Li et al. [25] reported a diode-side-pumped simultaneous dual-wavelength Nd:YAG laser at 1116 and 1123 nm.

Simultaneous generation of dual-wavelength have been obtained in many crystals, such as, Nd:YAP [26], Nd:YVO$_4$ [27], Nd:GdVO$_4$ [28], Nd:CNGG [29], Nd:APLN [30], Yb:YAG [31], V:YAG [32], Cr:YAG [33] and, etc.

Especially, the most investigation on Nd:YAG systems concentrated on 1319/1318 nm, 1052/1064 nm, 1.06/1.3 μm [34], 1318/1338 nm [35], 946/938 nm and 946/1064 nm [36-38] lines.

In present research simultaneous oscillation of dual-wavelength Nd:YAG laser at 938 and 946 nm pumped by flashlamp was introduced for the first time.

2.2.3. Background of Effects of Temperature on Laser Performance

Following the classical paper by McCumber [39] in which they studied on the shift and bandwidth of the R-lines in ruby, many research works have been carried out on thermally induced line broadening in trivalent rare earth ions doped in different crystal hosts [40-45].

Kushida [46] measured linewidth and thermal shifts of several lines in Nd^{3+}:YAG crystal. Similar work also reported by Xing group [47]. They have made a considerable effort on the shifts of the all spectral lines in the manifold of $^4F_{3/2}$ to $^4I_{11/2}$. Furthermore, they also studied the shift changes between the Stark sublevels R_1 and R_2 of Nd^{3+}:YAG crystal as a function of temperature over the range of 20 to 200°C. Optical parameters of energy level and temperature dependent of transition lines of Nd^{3+} in $CaZ_2Y_2Ge_3O_{12}$ was characterized [48].

The thermal effects on the bandwidths, positions and line shifts of 933.6 and 1057.3 nm lines of Nd^{3+} doped lanthanum lutetium gallium garnet (LLGG) crystal host within the respective intermanifold transitions of $^4I_{3/2}$ to $^4I_{9/2}$ and $^4I_{11/2}$ was investigated by Sardar et. al. [49]. In addition, they investigated the emission spectra for the transitions from intermanifold $^4F_{3/2}$ to $^4I_{9/2}$ and $^4I_{11/2}$ of Nd^{3+} in YVO_4 crystal host. Besides effects of temperature on the position and width of the main laser line of Nd^{3+}: YVO_4 at 1061.7 nm was also examined by this group. Mao et al. measured the absorption and emission lifetime of upper state level of Nd^{3+} ions in YVO_4 (2 at %) laser host at different temperature [50].

Spectroscopic properties and stimulated emission cross section of Nd:YAG crystals at room temperature have been studied extensively [51-55]. Similar studies also have been reported by Rapaport group who only focused the study on the major line of the 1064 nm emission cross section and a lifetime of the Nd^{3+} doped YAG crystal over the range of temperatures from -70°C to +70°C [56, 57]. Furthermore, Dong et al. investigated the absorption and emission spectra, stimulated emission cross section, and radiative lifetime of Nd:YAG crystals doping at various concentration including 1, 2 and 3 at % Nd^{3+} at different temperature in the range between 70 to 300 K.

The temperature dependence of long-pulse and Q-switched operation of Nd^{3+} doped solid-state lasers were investigated for 1064 nm wavelength [58]. The effects of temperature on the laser threshold, laser output energy and slope efficiency in the range of -60 to 60 C was described. This work is useful to design Q-switched and free running solid state lasers to operate at different temperatures without damaging intracavity optics.

However, to the best of our knowledge, input energy dependency of linewidth and wavelength position of quasi-three-level and four level system transitions which were investigated at the present work has not been reported up to now. In addition, laser performance of quasi-three-level transitions at 938 and 946 nm versus temperature and input energy was studied.

2.3. THEORY OF THERMAL EFFECTS ON LASER PERFORMANCE

2.3.1. Introduction

Optically elements in a typical laser resonator (e.g., mirrors, beam splitters, etc.) show no variation of optical properties over a wide range of temperature. However, some parameters such as laser threshold, output power, internal loss, the linewidth of the laser line and so on are dependent on temperature. Therefore heat generation can dramatically influence the laser performance. Furthermore, in most high power solid state lasers variation of emission cross section with temperature has a serious effect on the laser performance.

This chapter aims to derive some useful equations for stimulated emission cross-section. Furthermore, main mechanisms theory of the spectral line thermal broadening and thermal shifts is investigated. Finally, Three, four and quasi-three-level laser systems were investigated.

2.3.2. Einstein's Approach to Blackbody Radiation Problem

In 1917, three kinds of transition rates (variation in the number of atoms per unit time, *dN/dt*), was identified by Einstein, i.e., the spontaneous emission rate (A_{21}), the stimulated emission rate (B_{21}), and the induced absorption rate (B_{12}). These rates are important to calculate of pumping requirements and laser gain.

In equilibrium, there must be a detailed balance between processes transferring atoms from state 1 to state 2 and processes transferring atoms from state 2 to state 1 (Figure 2.1). The detailed balance equation can be expressed as

$$g_1 B_{12} e^{(-h\nu_1/k_B T)} U(\nu) = g_2 A_{21} e^{(-h\nu_1/k_B T)} + g_2 B_{21} e^{(-h\nu_2/k_B T)} U(\nu) \quad (2.1)$$

The equation at the left side shows induced absorption. It includes the product of the degeneracy of the lower state (g_1), the stimulated rate of electrons that go from the lower state to upper state (B_{12}), the radiative probability of the lower state ($e^{(-h\nu_1/k_B T)}$), and stimulating radiation field ($U(\nu)$).

The first term of right side of equation represents spontaneous emission from the upper state which includes the product of the degeneracy of the upper state g_2, the spontaneous rate of electrons that leave the upper state and go to the lower state (A_{21}), and relative probability of the upper state ($e^{(-h\nu_2/k_B T)}$). The second term shows the stimulated emission from the upper state which comprise the product of the degeneracy of the upper state (g_2), the stimulated rate of electrons that leave the upper state and go to the lower state (B_{21}), the relative probability of the upper state ($e^{(-h\nu_2/k_B T)}$), and the stimulating radiation field ($U(\nu)$).

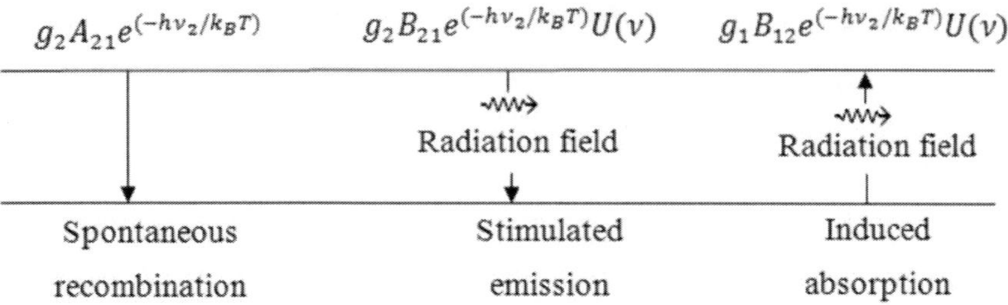

Figure 2.1. Balance between processes transferring atoms from state 1 to state 2 and from state 2 to state 1 in equilibrium.

2.3.3. Stimulated Emission Cross Section

Three basic interaction namely absorption, spontaneous and stimulated emissions would occur in a laser medium when an electromagnetic wave interacts with a material [As explained in section 2.3.2].

Atomic systems containing atoms, molecules, and ions can be solely at distinct energy levels. A variation from one energy level to other, known as a transition and causes either emission or absorption of a photon. The frequency of the absorbed or emitted transition is given by

$$E_2 - E_1 = h\nu_{21} \tag{2.2}$$

where E_1, E_2 and ν_{21} are two different energy states, and the frequency of the transition respectively.

In thermal equilibrium, the relative populations of any two energy levels E_1 and E_2 at temperature T can be estimated By the Boltzmann ratio

$$\frac{N_2}{N_1} = \frac{g_2}{g_1} \exp\left(-\frac{E_2 - E_1}{kT}\right) \tag{2.3}$$

where N_1 and N_2 are the numbers of atoms in the energy states of E_1 and E_2, respectively. T is temperature, $k = 1.38 \times 10^{-16}$ erg.K^{-1} is Boltzmann's constant, and $g_i (i = 1,2)$ is the degeneracy of energy levels.

The lower energy levels in the medium are more considerably populated than the upper energy levels while the system is in thermal equilibrium [59]. However, if the atom is disturbed by an electromagnetic radiation with frequency $v = v_{21}$, the atom will be excited to level 2 and therefore absorption occurs. The return of the system to initial equilibrium, i.e., the 2→1 transition, can occur through two processes i.e., spontaneous (radiative) emission and stimulated emission (eq. 2.1). The condition for equilibrium of the system by using Eq. (2.1), expression (2.3) takes the form is

$$U(v_{12})(B_{12}\exp(hv_{21}/kT) - B_{21}) = A_{21} \qquad (2.4)$$

Hence, a simple relationship is derived between the spectral flux density and the Einstein coefficients

$$U(v_{21}) = \frac{A_{21}}{B_{12}\exp(hv_{21}/kT) - B_{21}} \qquad (2.5)$$

Comparison of this expression with Plank formula gives

$$U(v) = \frac{8\pi h v^3}{c^3}\frac{1}{\exp(hv/kT) - 1} \qquad (2.6)$$

Readily shows a relation between the Einstein coefficient (when the medium is not a vacuum),

$$B_{21} = B_{12} \qquad (2.7)$$

     Seyed Ebrahim Pourmand, Iraj Sadegh Amiri et al.

$$B_{21} = \frac{c^3}{8\pi h\, v_{21}^3 n^3} A_{21} \tag{2.8}$$

Or generally

$$B_{ij} g_i = B_{ji} g_j \tag{2.9}$$

$$B_{ji} = \frac{c^3}{8\pi h\, v_{ji}^3 n^3} A_{ji} \tag{2.10}$$

where c and n are the light velocities in the vacuum and the refractive index of the material. The factor before A_{21} represents the number of radiative oscillators (oscillation modes) per unit volume emitting over a single frequency range.

In considering the main laser properties, it is better to characterize induced radiative transitions by a parameter called the cross-section, σ_e, rather than by probability. If an emission spectrum consists of a single narrow line with the form factor $g(v)$, the parameter σ_e can have various forms. We shall use

$$\sigma_e = B_{ji} \frac{h\, v_{ij}\, n}{c} g(v) = A_{ji} \frac{c^2}{8\pi v_{ij}^2 n^2} g(v) \tag{2.11}$$

The transition line shapes in Nd:YAG is fitted by Lorentzian so we have

$$g(v) = \frac{2}{\pi \Delta v} \tag{2.12}$$

Hence

$$\sigma(i \to j) = \frac{\lambda_{ij}^2}{4\pi^2 n^2 c \Delta v} A(i \to j) \tag{2.13}$$

Here λ_{ij} and Δv are the wavelength and bandwidth (FWHM) of emission line respectively. Expression (2.13) is one of the various forms of the well-known Fuchtbauer-Ladenburg formula.

For a Gaussian luminescence line shape and emission, cross-section can be reprinted as

$$g(v) = \frac{2}{\Delta v} \sqrt{\frac{\ln 2}{\pi}} \tag{2.14}$$

$$\sigma(i \to j) = \frac{\lambda_{ij}^2}{8\pi n^2 c \Delta v} A(i \to j) \tag{2.15}$$

The radiative probability of the inter-Stark transition $A(i{\to}j)$ for Nd:YAG can be expressed as [60, 61]

$$A(i \to j) = \left(1 + e^{-\Delta/kT}\right)\beta(i \to j) \times \beta(I \to K)\tau(^4F_{3/2})^{-1} \tag{2.16}$$

where Δ is the difference in energy between the Stark levels R_1 and R_2 of the $^4F_{3/2}$ manifold, k is the Boltzmann constant, T is the absolute temperature, $\beta(i{\to}j)$ represents the branching ratio for the corresponding inter-Stark transition, and $\beta(I{\to}K)$ is the branching ratio for the respective intermanifold transition. The branching ratio for the inter-Stark transition is determined by comparing the luminescence density of the inter-Stark transition to the total luminescence density for transitions that terminate at the Stark levels of the $^4I_{9/2}$ manifold.

2.3.4. The Homogeneously Broadened Line Shape

The main difference between homogeneously and inhomogeneously broadened emission lines appears in the saturation action of these emission lines and as a result, an extensive effect on the laser performance will occur. However homogeneous line shape will saturate without variation by an adequately powerful signal utilized inside the atomic linewidth. Lifetime and thermal broadening are the mechanisms which cause homogeneously broadened line shapes.

2.3.5. Lifetime Broadening

Lifetime broadening is due to the decay mechanisms of the atomic system. Electrons transit to another energy level, shorten the lifetime and so broadening the energy level. Radiative lifetime of spontaneous emission is intrinsic because this emission is an inevitable feature of any transition. Broadening of the atomic transition is caused by spontaneous emission and is obtained by uncertainty relation $\Delta E \Delta t = \hbar$ and the energy is expressed in the unit of wave number according to 1 erg. $= 5.035 \times 10^{15}$ cm^{-1}. Actually, this kind of effect is not the main contribution of the electron-phonon interaction on the line width of zero phonon lines. One can see it clearly from the example of the pure electron transition line R_1 of the trivalent chromium in ruby and Nd:YAG crystal.

Actually, the lineshape and linewidth due to the spontaneous emission process itself are intensively small. For instance in Nd:YAG laser the upper-level energy decay is τ=230 µs and thus the contribution of lifetime broadening is only 700 *Hz*, which is absolutely insignificant compared to the enormously larger phonon broadening dephasing contribution which is around 120 *GHz* [62].

However another mechanism such as the single phonon absorption and emission as well as the Raman scattering, within the lifetime of the electron energy levels, have their main contribution to the spectral linewidth.

2.3.6. Thermal Broadening

Thermal broadening is caused by interaction between impurity ions with lattice vibrations in crystalline solids. At high frequencies, the ions are surrounded by the thermal vibrations of the lattice and then the atoms resonance frequency will modulate. This modulation of frequency results in coupling behavior between the atoms, and therefore a homogeneous linewidth is obtained. Linewidth broadening of the ruby (Cr^{3+}: Al_2O_5) and neodymium-doped in yttrium aluminum garnet (Nd^{3+}: YAG) is caused by thermal broadening.

However, each homogenous broadening spectral line has Lorentz line shape and its line shape function for the normalized Lorentz distribution is given by

$$g(v) = \left(\frac{\Delta v}{2\pi}\right)\left[(v-v_0)^2 + \left(\frac{\Delta v}{2}\right)^2\right]^{-1} \tag{2.17}$$

Here, v_0 and Δv are the center frequency and the full width at half maximum of amplitude (FWHM) respectively. For the Lorentzian line shape, the peak value is

$$g(v_0) = \frac{2}{\pi \Delta v} \tag{2.18}$$

The effects of temperature online shape function of crystals will discuss in the following.

2.3.7. The Inhomogeneously Broadened Line Shape

The inhomogeneous broadening emerges when the local electric field is different from ion to ion and so the Stark effect changes the energy levels in an inhomogeneous way. Random crystalline defects may bring about inhomogeneously broadened Solid-state lasers. This occurs only at low temperatures where the lattice vibrations are small. Random variations of lattice strains and displacements may result in negligible shifts in the precise energy states and emission line wavelengths from ion to ion.

Line shape of the inhomogeneously broadened bandwidth is expressed by a Gaussian function as below

$$g(v) = \frac{2}{\Delta v}\left(\frac{\ln 2}{\pi}\right)^{1/2} \exp\left[-\left(\frac{v - v_0}{\Delta v/2}\right)^2 \ln 2\right] \qquad (2.19)$$

Peak value of the curve is

$$g(v_0) = \frac{2}{\Delta v}\left(\frac{\ln 2}{\pi}\right)^{1/2} \qquad (2.20)$$

2.3.8. Main Mechanism of the Spectral Line Thermal Broadening and Thermal Shifts

The spectral line width of the active ions is increased and their peak position is shifted when the temperature is raised. Beside the crystal field weakens by lattice thermal expansion, the phonon transition has its effect on the energy of electron-phonon system and modulation of Coulomb interaction between electrons and spin-orbital coupling. All of these have their contributions to the spectral thermal line shift.

In a pure and perfect crystal, these phenomena mainly come from the effect of electron-phonon interaction. This effect causes the shift and broadening of the initial and final electron states, and so it is the starting point of the study of spectral line thermal broadening and thermal shift.

Many theoretical models have been introduced to explain thermal broaden and shift to spectral line. However the model proposed by McCumber is highly regarded as enormously influential work, which has brevity and clarity, and its main conclusion can be applied quite generally. Therefore we will use this model in the discussion of line width broaden and line shift, but introduce the modification of mass difference of the host ions. It should be pointed out that all the different models published so far contain a series of major simplification of the real physical reality (for example to use long wavelength approximation of the lattice vibration and the Debye distribution of phonon frequency), and so can only be used to describe approximately the related phenomena.

2.3.9. The Contribution of Single Phonon Absorption and Emission to the Spectral Linewidth

In present work, just practical equations were highlighted. To see derivation and details of equations please refer to books which were addressed in references (Powel 1998; Luo 2007; Bartolo 2010).

To calculate the transition rates of the single phonon absorption and emission, one should, at first, expand the energy of the electron-phonon interaction in terms of strain. The crystal strain is a tensor obviously. However, in the study of thermal vibration effect on the spectral line width, the direction problems can be neglected and so the displacement of the lattice can express by a scalar quantity u [63]

$$u = \frac{1}{\sqrt{m}} \sum_k Q_k e^{ikR} \qquad (2.21)$$

where m = NM_α is the mass of the crystal, Q_k is the normal coordinates. For acoustic mode k = ω_k / v.

Varying in the crystal field owing to fluctuating inactive ions location and their encircling ligands causes the electron-phonon interaction. This is proportionate to average of local strain which defined as

$$\varepsilon \approx \left. \frac{\partial u}{\partial R} \right|_{R=0} = i \sum_k \left(\frac{\hbar \omega_k}{2 M v^2} \right)^{1/2} \left[a(k) - a^+(k) \right] \qquad (2.22)$$

where $a(k)$ and $a^+(k)$ are annihilation and creation operators respectively.

Expanding the energy of the crystal field in terms of the local strain is

$$V_{e-p} = V_0 + V_1 \varepsilon + V_2 \varepsilon^2 + \dots \qquad (2.23)$$

where V_0 is crystal field and the remaining terms in the right hand represent the electron-phonon interaction.

The linear term of ε in Eq. (3.35) will be

$$H_{e-sp} = i V_1 D \sum_k \left(\frac{\hbar \omega_k}{2 M v^2} \right)^{1/2} \left[a(k) - a^+(k) \right] \qquad (2.24)$$

Eq. (2.24) is the Hamiltonian of single phonon transition. According to Fermi golden rule, the transition rate for single phonon absorption or emission will be obtained. However, these kinds of phonon absorption and emission are reversible and it does not change the lifetime of electron energy levels. In addition, a broadening of the initial state i resulted from the phonon absorption and emission will be $\Delta E_{sp}^{a} = W_{sp}^{a} h$ and $\Delta E_{sp}^{e} = W_{sp}^{e} h$ respectively.

In the energy unit of wave number cm^{-1}, 1 erg = 5.035 × 10^{15} cm^{-1} then h = 6.62620 × 10^{-27}erg.sec = 3.3362917 × 10^{-11} cm^{-1}.sec = 1/c. Therefore

$$\Delta E_{sp}^{a}(cm^{-1}) = \sum_{f\rangle i} \frac{1}{c} \frac{\left(\omega_{fi}^{e}\right)^{3} D^{2}}{2\pi\rho\hbar} \left(\frac{1}{v_{l}^{5}} + \frac{2}{v_{t}^{5}}\right) \left|\langle f|V_{1}|i\rangle\right|^{2} \left(\frac{1}{e^{\hbar\omega_{fi}^{e}/kT} - 1}\right) \quad (2.25)$$

$$\Delta E_{sp}^{e}(cm^{-1}) = \sum_{f\langle i} \frac{1}{c} \frac{\left(\omega_{fi}^{e}\right)^{3} D^{2}}{2\pi\rho\hbar} \left(\frac{1}{v_{l}^{5}} + \frac{2}{v_{t}^{5}}\right) \left|\langle f|V_{1}|i\rangle\right|^{2} \left(1 + \frac{1}{e^{\hbar\omega_{fi}^{e}/kT} - 1}\right) \quad (2.26)$$

β_{fi} is usually in literature to denote the product of the physical quantities in above formulae

$$\beta_{fi} = \frac{1}{c} \frac{\left(\omega_{fi}^{e}\right)^{3} D^{2}}{2\pi\rho\hbar} \left(\frac{1}{v_{l}^{5}} + \frac{2}{v_{t}^{5}}\right) \left|\langle f|V_{1}\rangle i\rangle\right|^{2} \quad (2.27)$$

Neglecting the difference between the velocity of transverse and longitude acoustic wave, above formula can be written as

$$\beta_{fi} = \frac{3}{c} \frac{\left(\omega_{fi}^{e}\right)^{3} D^{2}}{2\pi\rho\hbar v^{5}} \left|\langle f|V_{1}\rangle i\rangle\right|^{2} \quad (2.28)$$

Hence

$$\Delta E_{sp}^{a}(cm^{-1}) = \sum_{f\rangle i} \beta_{fi} \frac{1}{e^{\hbar\omega_{fi}^{e}/kT} - 1} + \sum_{f\langle i} \beta_{if} \frac{1}{e^{\hbar\omega_{if}^{e}/kT} - 1} + \sum_{f\langle i} \beta_{if} \quad (2.29)$$

where the first term is the contribution of the phonon absorption and the following two terms are those of the phonon emission.

At low temperature, two $\dfrac{1}{e^{\hbar\omega^e_{fi}/KT}-1}$ factors in the summation of expression (2.29) are much smaller than 1 and so the main contribution comes from the third term.

2.3.10. The Contribution of Phonon Raman Scattering to the Spectral Linewidth

The perturbation theory will be used to introduce the related formulae. The reader can refer to the paper published by Skinner and Hsu to know the method of the non-perturbation theory [64]. Raman scattering is a second perturbation process, it contains second-order perturbation of the first order perturbation of the following order term

$$H_{e-dp} = -V_2 \frac{\hbar D^2}{2Mv^2} \sum_{kk'} \sqrt{\omega\omega'} \left(a(k)-a^+(k)\right)\left(a(k)-a^+(k)\right) \quad (2.30)$$

Similar to equation (2.24), the transition rate of the Raman scattering can be calculated by

$$W_R = \frac{2\pi}{\hbar^2} \int \left|\langle f|H_R|i\rangle\right|^2 \rho(\omega_f)d\omega_f \quad (2.31)$$

The electronic state should be the same state denoted by α but with different phonon state for the initial and final states of the Raman scattering process discussed.

Derivation of Equations (2.30) and (2.31) can lead to the formula of linewidth as a function of temperature

$$\Delta E(T)(cm^{-1}) = \overline{\alpha}\left(\frac{T}{T_D}\right)^7 \int_0^{T_D/T} \frac{x^6 e^x}{(e^x - 1)^2} dx \equiv \overline{\alpha} F\left(\frac{T}{T_D}\right) \qquad (2.32)$$

where

$$\overline{\alpha} = \frac{1}{c}\frac{9D^4}{2\pi^3 \rho^2 v^{10}}\left(\frac{kT_D}{\hbar}\right)\left(\sum_{\gamma \neq \alpha} \frac{\left|\langle\alpha|V_1|\gamma\rangle\right|}{E_\alpha - E_\gamma} + \langle\alpha|V_2|\alpha\rangle\right)^2 \qquad (2.33)$$

where $\frac{\overline{\alpha}}{T_D^7}$ is a parameter depending on the density of the crystal, the phonon velocity and electron-phonon coupling strength and is actually adjustable. Its value is proportional to D^4, and the ion mass difference of the host has a much obvious effect on the line thermal broaden than that in the case of phonon emission and absorption mechanisms. The range of $\overline{\alpha}$ is $\overline{\alpha} \leq 100 cm^{-1}$ for rare earth ions and $\overline{\alpha} \geq 300 cm^{-1}$ for transition metal ions.

The above-mentioned mechanisms have different contribution at a different temperature. At very low temperature, the single phonon emission and absorption have the major contribution and that from Raman scattering is small. At higher temperature, the main contributions from Raman scattering, which covers the phonon spectrum and is independent of the energy separation of the light transition.

It should be pointed out that the thermal linewidth broadening produced by electron-phonon interaction belongs to the so-called homogeneous broadening. In a crystal with considerable defects, the linewidth broadening of the active ions often dominated by inhomogeneous broadening. In this kind of crystal, different ions can be in different positions with different strain, different lattice distortion, and so different crystal field interaction.

The spectrum observed is actually a superposition of the different spectra produced by the ions in different positions. Due to the random distributions of the strain, defects and impurity ions, the inhomogeneous broadening spectral line has a Gaussian line shape and each homogenous broadening spectral line has a Lorentz line shape.

In the general case, the line shape of the spectral line is a Voigt line shape, which is a convolution integral of the Lorentz lines with a Gaussian frequency distribution. When the width of the Gaussian distribution is zero (in the case of an ideal crystal without inhomogeneous), the line shape becomes Lorentz. On the other hand, when the width of Lorentz line tends to zero, the spectral line takes the shape of Gaussian. That is the line shape at very low temperature, in which the effect of lattice vibration is very weak. At high temperature, the line width of the Lorentz components can be much larger than that of the Gaussian in the composite Voigt line and so the integral lineshape is practically Lorentz. This is the reason why the spectral lines of the active ions at room temperature in the reasonable perfect crystal can be described by Lorentz line shape.

2.3.11. Calculation of the Thermal of Spectral Lines

To discuss thermal shifts of the spectral lines, it is natural to think that the thermal expansion is the main mechanism because thermal expansion will weaken the crystal field and shorten the energy level separations. Certainly, for the so-called "soft crystal" having large thermal expansion coefficient, this mechanism has the major contribution. However, it is not true for the "hard crystal" like YAG crystal. Actually, the calculation results according to the mechanism of thermal expansion cannot account for the experimental data of thermal shifts observed, even obtain a result with a different sign in some situations. Therefore, the electron-phonon interaction can be considered as the major mechanisms of the thermal shifts of spectral lines. Owing to realize the relation between the thermal shift of the spectral line and the temperature, it is enough to calculate the diagonal matrix elements by a general method. Expanding the electron-phonon interaction energy in terms of the normal coordinates of lattice vibration to second-order term [65]

$$H_{e-p} = H_{e-sp} + H_{e-dp} \qquad (2.34)$$

where H_{e-sp} and H_{e-dp} present the single phonon and two phonon term. The thermal shift in the unit of wave number can be expressed

$$\delta(E) = \alpha \left(\frac{T}{T_D} \right)^4 \int_0^{\frac{T_D}{T}} \frac{x^3}{e^x - 1} dx \qquad (2.35)$$

where α can be expressed by Eq. (3.72)

$$\alpha = \frac{3V}{4\pi^3 v^3 c} \left(\frac{kT_D}{\hbar} \right) a \qquad (2.36)$$

Thermal of shift lines can be expressed in the unit of wave number

$$\delta E_i (cm^{-1}) = \sum_{\beta \neq \alpha} \frac{1}{2\pi^2} \beta_{\alpha\beta} \left(\frac{T}{T_{\alpha\beta}} \right)^2 P \int_0^{\frac{T_D}{T}} \frac{x^3}{e^x - 1} \frac{1}{\left(T_{\alpha\beta}/T \right)^2} dx \qquad (2.37)$$

The temperature dependent changes of widths and positions of sharp spectral lines in crystals have been investigated by many researchers on 3d, 4f, and 5d ions. They mainly explain the relation to simple ion-phonon interaction as a perturbation and of a Debye phonon distribution. The ion-phonon interaction that causes the thermal broadenings and shifts of the energy levels of the impurity ion is worth looking [66].

A famed equation for the temperature dependence of the ith energy level width is given by:

$$\Delta E_i (cm^{-1}) = \Delta E_i^{strain} + \Delta E_i^{D} + \Delta E_i^{M} + \Delta E_i^{R} \tag{2.38}$$

where the terms on the right-hand side of the equation represents the broadenings due to the crystal strain inhomogeneity (ΔE_i^{strain}), direct one phonon processes (ΔE_i^{D}) between the i-th energy state and other close states (j), multiphonon decay processes (ΔE_i^{M}) which occur when energy distinction of two states are larger than the largest energy of the existent phonons and are essentially temperature independent, and the Raman phonon scattering processes (ΔE_i^{R}) which is temperature dependent and related to phonon scattering by the impurity ions.

The direct one-phonon broadening processes, comprise of a temperature dependent part, ΔE_i^{DT}, and a temperature independent part, $\sum_{j<i} \beta_{ij}$. Temperature independent part is due to spontaneous one-phonon emission and since the energy separation between the Stark levels is of the order of $10\text{-}10^2 cm^{-1}$ for rare earth ions, $\sum_{j<i} \beta_{ij}$ can cause an observable broadening. Equation 1 can be written as:

$$\Delta E_i (T) = \Delta E_i^{strain} + \sum_{j<i} \beta_{ij} + \Delta E_i^{D}(T) + \Delta E_i^{M} + \Delta E_i^{R}(T) \tag{2.39}$$

Even at the temperature of $T = 0$ K, the energy level broadening is emerged due to residual width of the ith level, ΔE_{i0}, which is produced by crystal strains, one-phonon and multiphonon emission processes. Equation (3.79) yields:

$$\Delta E_i = \Delta E_i^{strain} + \Delta E_i^{M} + \sum_{j<i} \beta_{ij} \equiv \Delta E_{i0} \tag{2.40}$$

Inhomogeneous broadening due to crystal strains gives rise Gaussian transition line shape while one phonon emission process, multiphonon emission process and the Raman scattering process lead to homogeneous broadening with Lorentzian transition lineshape.

Total linewidth relevant to the two energy levels is expressed as:

$$\Delta E(cm^{-1}) = \Delta E_i(cm^{-1}) + \Delta E_f(cm^{-1}) \tag{2.41}$$

where E_i and E_f represent the initial energy level and the terminal level of the optical transitions respectively.

2.4. THREE, FOUR AND QUASI THREE LEVEL LASER SYSTEMS

Lasers are classified according to the number of energy states available in laser operation. Actually, there are three groups of laser systems which are three, four and quasi-three-level systems. In an ordinary two-level system is not generally practical for laser action. Because the particles pumped into the upper laser state has an equal probability of stimulating them to return to lower state. In addition, the upper laser state must have sufficient lifetime so that the particles live long enough to be stimulated and thus contribute to the laser gain [67].

However, the ground state is almost completely filled at thermal equilibrium. When the population of the atoms at the lower state exceeds that of the upper state, emitted photons of the laser light are actually absorbed and laser action is not possible.

Therefore, a higher rate of the stimulated emission than absorption is required (Csele, 2004). When the numbers of atoms in both states are equal, the rate of stimulated emissions becomes equal to the rate of absorption, the active material is then transparent to the incident radiation and emission processes are the majority and the radiation is enhanced (Koechner, 2006).

In addition, a mechanism where $N_2 > N_1$ must be obtained and this is called population inversion. A population inversion can be created by introducing a metastable level where atoms can fill up to achieve a situation where more N_2 than N_1.

2.4.1. Three-Level Laser System

In three-level systems, the laser transition terminates on the ground level. The atoms or ions are pumped from the ground state E_1 to higher state E_3 (pumping state) by using pumping source. E_3 is unstable state and here atoms stay for 10^{-9} to 10^{-8} seconds.

The majority of the excited atoms are transferred quickly as nonradiative transitions to the upper laser state E_2 generally, generates heat in the material. Energy level E_2 is the metastable state (having lifetime 10^{-5} to 10^{-3} seconds). Thus, the population of atoms becomes more in the energy state E_2 as compared to E_1. Finally, the stimulated emission occurs between E_2 and E_1 producing laser decay.

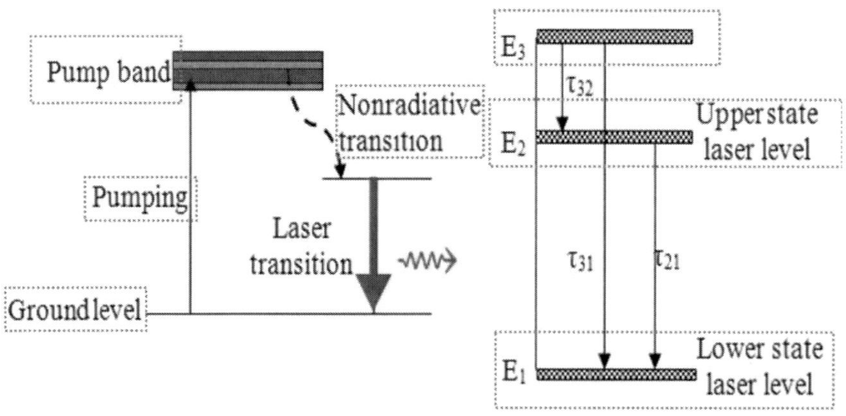

Figure 2.2. Schematic diagram of the three-level laser system.

Since in a three-level laser system, the terminal level is strongly populated, level 2 must be populated more than half of the atoms in the ground state so that the lasing action can emerge. For instance, ruby (Cr^{3+}: Al_2O_3) is a three-level laser medium. Figure 2.2 indicates the schematic diagram of the three-level systems.

If pumping intensity is below laser threshold, atoms in level 2 return to ground level by spontaneous emission. As the terminal level of the laser transition is the highly populated, a very high population must be reached in level 2 before population inversion occurred so that the lasing action can start. The drawback of the three-level system is that more than half of the atoms in the ground state must be lifted up to the excited state E_2. This requires very strong pumping and leads to a low efficiency. The four-level systems avoid this limitation.

2.4.2. Four-Level Laser System

Figure 2.3 shows the schematic diagram of the four-level laser systems. In this system, optical pumping excites the atoms from the ground state (E_1) into the pump band (E_4). From this level, the excited atoms will rapidly decay to the metastable level E_3 by the non-radiative transition.

However, the lifetime relaxation of excited atoms in E_3 level is sufficiently long compared to E_4 and E_2 levels thus a population accumulates in this level 3 (upper lasing level). Then the atoms proceed to terminal level E_2 (lower laser level) to create laser transitions through spontaneous and stimulated emissions. The lower laser state E_1 is located above the ground state, therefore, its thermal population is negligible. Finally, the atoms transfer quickly to the ground level and produce non-radiative decays.

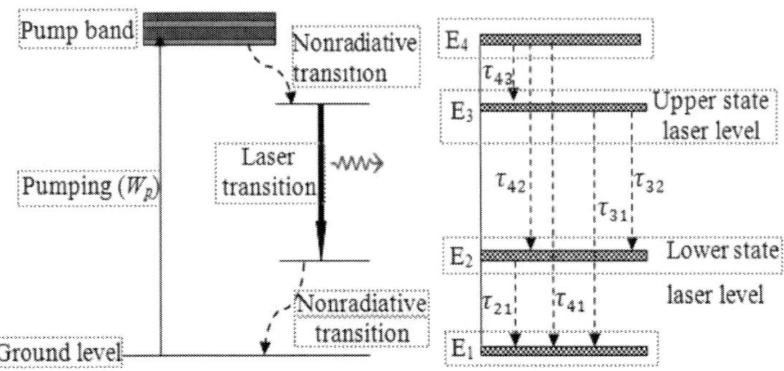

Figure 2.3. Schematic diagram of the four-level laser system.

Since only a small number of atoms need to be excited in the upper lasing level E_3 to form population inversion, so the threshold can be obtained easily in a four-level laser system compared to three level system [68].

2.4.3. Quasi Three-Level Laser System

A quasi-three-level laser medium is an intermediate between four-level and three-level systems. Figure 2.4 shows quasi-three-level laser system. In a quasi-three-level, the lower laser level is a sublevel of the ground level. Thus lower laser state has the significant thermal population and so laser output suffers from significant re-absorption.

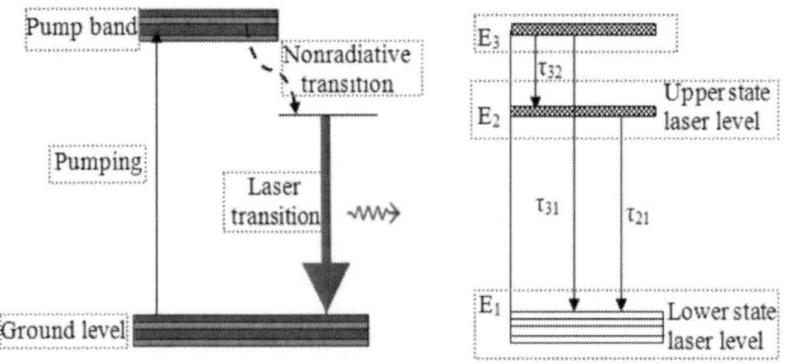

Figure 2.4. Schematic diagram of the quasi-three-level system.

This phenomenon increases the internal loss in the gain medium and therefore population inversion would emerge at strong pump intensities. As a result, additional heat is generated by the crystal. For instance Nd:YAG laser operation at 938 and 946 nm is a quasi-three-level laser.

2.5. THREE AND FOUR LEVEL LASER TRANSITION RATES AND THRESHOLD CONDITIONS

The properties of laser cavity are dependent on the spectroscopic characteristics of the laser host by knowledge of its amplification factor. Population inversion occurs when ions in the upper laser level are more than ions in the ground state. The amplification coefficient at the maximum amplitude of line shape (α_a), and small-signal gain ($g_0(\lambda)$) are determined as

$$\alpha_a = -\sigma_e \Delta n \tag{2.42}$$

$$g_0(\lambda) = \sigma_e(\lambda)\Delta n \tag{2.43}$$

$$\Delta n = n_2 - \frac{g_2}{g_1} n_1 . \tag{2.44}$$

where Δn and σ_e are the population inversion factor and the emission cross-section.

In order to sustain the lasing performance, the rate of the round-trip losses has to be smaller than round-trip gain in the resonator of the laser. The laser is at the threshold when the gain and loss are equal. For laser performance threshold can be obtained by considering the round trip gain in the cavity, containing losses. A round trip gain equates to one at the threshold, therefore [69]

$$G = R_1 R_2 \exp[(g - \alpha)2l] = 1 \tag{2.45}$$

where R_1 and R_2 are reflectivities of the cavity mirrors, g is the gain coefficient, α is the loss coefficient, and l is the length o the cavity. Then

$$2gl = T + L \tag{2.46}$$

where T and L stand for the output of active cavity and the passive losses of the cavity respectively,

$$T = -\ln R_{out} \tag{2.47}$$

where R_{out} is the reflectivity of the output coupler.

To calculate the gain coefficient, population inversion should be defined at the threshold for various pumping cases. The rate equations showing the time development of the ground and excited state populations for an optically pumped three level system are expressed by the diagram of Figure 2.2. In this case, generally, the relaxation rate from the pump band to the upper laser level is much greater than the rate of emission from the pump band to the ground state. In addition, the relaxation time from the pump band to upper laser level is rapidly compared to the spontaneous lifetime of the metastable level ($\tau_{32} \ll \tau_{31}$). Under these conditions, rate equation for three level system is written as [70]

$$\frac{dn_1}{dt} = \left(n_2 - \frac{g_2}{g_1}n_1\right)c\phi\sigma + \frac{n_2}{\tau_{21}} - W_p n_1 \tag{2.48}$$

And

$$\frac{dn_1}{dt} = -\frac{dn_2}{dt} \tag{2.49}$$

Here ϕ, σ and W_p are the density of photons in the cavity mode, the stimulated emission cross section between levels 2 and 1, and the pumping rate respectively.

Using Eq. (2.44) and substituting Eqs. (2.44) and (2.48) gives

$$\frac{d\Delta n}{dt} = -\left(1 + \frac{g_2}{g_1}\right)\Delta n \phi \sigma c - \frac{\Delta n + n_T \frac{g_2}{g_1}}{\tau_f} + W_p(n_T - \Delta n) \tag{2.50}$$

where $n_T = n_1 + n_2$ and $\tau_f = \tau_{12}$.

In addition to accepting of effective and rapidly relaxation from the pump band to the laser level, the same rate equation can be found for a four laser level system as

$$\frac{dn_2}{dt} = \left(n_3 - \frac{g_3}{g_2}n_2\right)c\phi\sigma + \frac{n_3}{\tau_{32}} - \frac{n_2}{\tau_{21}} \tag{2.51}$$

$$\frac{dn_3}{dt} = -\left(n_3 - \frac{g_3}{g_2}n_2\right)c\phi\sigma + \frac{n_3}{\tau_{32} + \tau_{31}} + W_p n_1 \tag{2.52}$$

For standard conditions $\tau_{21} \approx 0$ and $n_T = n_1 + n_3$ so $\Delta n = n_3$. Thus

$$\frac{d\Delta n}{dt} = -\Delta n \phi \sigma c - \frac{\Delta n}{\tau_f} + W_p(n_T - \Delta n) \tag{2.53}$$

For both four and three-level systems, the phonon density in the cavity is described by

$$\frac{d\phi}{dt} = \phi\sigma c\Delta n - \frac{\phi}{\tau_c} + S \tag{2.54}$$

where τ_c and S are the photon lifetime and the rate of spontaneous emission that usually is negligible.

Since at threshold and above it, the photon density grows, therefore $d\phi/dt \geq 0$ and from Eq. (2.54) the population inversion is

$$\Delta n_{th} = \frac{1}{c\sigma\tau_c} \tag{2.55}$$

Threshold conditions of $\phi = 0$ and $d\Delta n/dt=0$ for the continuous wave, the fractional population inversion for three and four level systems are obtained from Eq. (2.50) and (2.53) to be

$$\frac{\Delta n_{th}}{n_T} = \frac{W_p\tau_f - \dfrac{g_2}{g_1}}{W_p\tau_f + 1} \text{ , three level system} \tag{2.56}$$

$$= \frac{W_p\tau_f}{W_p\tau_f + 1} \text{ , four-level system} \tag{2.57}$$

When pumping rates are small, $W_p\tau_f \ll 1$, for any pump rate there is a population inversion in a four level laser system, but the pump rate at threshold in a three laser level system is expressed as

$$W_p(th) = \frac{g_2}{g_1 \tau_f} \tag{2.58}$$

At threshold when $n_2 \ll n_1$ for a four level system, rate of fluorescence decay is much greater than pumping rate, $W_p(th) \ll 1/\tau_f$ but Eq. (2.58) shows for a three level system that the level populations are closely equal, the pump rate at threshold is approximately equated to the spontaneous emission rate by considering the degeneracy factor.

However owing to stimulated emission, the photon density of the cavity mode is enormous at above threshold. The population inversion between the pumping rate and total spontaneous and stimulated emission rates is balanced when steady-state conditions are reached.

Small-signal gain is solely related to medium characteristics and the pump rate. In addition saturation gain also depends on the density of photons in the cavity mode. The density of the beam power is yield by $I = ch\nu\phi$, so the gain is expressed as

$$g = \frac{g_0}{1 + \dfrac{I}{I_s}} \tag{2.59}$$

where the saturated intensity is defined while the gain is one-half of the small-signal gain therefore

$$I_s = \left(W_p + \frac{1}{\tau_f}\right) \frac{h\nu}{\sigma\left(1 + \dfrac{g_2}{g_1}\right)} \ , \ \text{three-level system} \tag{2.60}$$

$$= \left(W_p + \frac{1}{\tau_f}\right) \frac{h\nu}{\sigma} \ , \ \text{four-level system} \tag{2.61}$$

The saturated intensity of a four-level system with $W_p \ll \tau_f^{-1}$ is written as

$$I_s = \frac{h\nu}{\sigma\tau_f} \tag{2.62}$$

The fluorescence power is written as

$$P_f = \eta_1\eta_2\,\eta_3\eta_4 P_{in} \tag{2.63}$$

where P_{in}, η_1, η_2, η_3, η_4 are the electrical input power, the pump efficiency containing the quantum defect, the pump source efficiency, the efficiency of coupling the pump light into the gain medium, and the power absorption efficiency respectively.

The small-signal gain coefficient for ideal four-level systems regarding the electrical input power at threshold,

$$g_0 = \frac{\sigma\tau_f\eta_1\eta_2\eta_3\eta_4 P_{in}(th)}{h\nu V} \tag{2.64}$$

where $V = Al$ is the capacity of the cavity with cross section A and length l. The cavity gain is defined as

$$\ln G = g_0 l = \theta P_{in}(th) \tag{2.65}$$

Eq. (2.62) for the saturation intensity, gives the efficiency θ for a four-level system

$$\theta = \frac{\eta_1\eta_2\eta_3\eta_4\eta_5}{I_s A} \tag{2.66}$$

Using Eqs. (2.45) and (2.47) threshold input power can be expressed in terms of the efficiency parameter θ and cavity losses δ as

$$P_{in}(th) = \frac{\delta - \ln R_{10}}{2\theta} \qquad (2.67)$$

By comparing Eq. (3.125) with experimental results, the laser characteristics can be obtained.

2.6. QUASI THREE LEVEL ND:YAG LASER PUMPED BY FLASHLAMP

In this section a theoretical model for quasi-three-level Nd:YAG laser parameters pumped by flashlamp is outlined. The populations of the upper and lower laser levels are expressed by presenting the fractional population coefficient $f_{a,b}$ for the two laser levels as $N_2=f_b N_u$ and $N_1=f_a N_0$, , where N_u and N_0 are the total populations of the upper and lower manifolds Figure 2.4. To find the rate equation of the population inversion ($N=N_2 - N_1$), It is assumed that there is not any ground level depletion population, ($N_u \langle\langle N_0$ and $N_0 \approx N_{0g}$ where N_g is the total dopant concentration of Nd^+-ions), which in the case of $f_a \langle\langle f_b$ is valid [71].

$$\frac{dN}{dt} = (f_a + f_b)W_p N_g - (f_a + f_b)Nc\sigma\phi - \frac{N - N^0}{\tau} \qquad (2.68)$$

where σ, c, ϕ, τ and W_p are the transition cross section for $\lambda = 946$ nm, the light velocity in the material, and photon density in the resonator, a lifetime of the upper laser level respectively. $N^0 = N_0^2 - N_0^1$ is the difference of equilibrium population between the two laser states.

Change of the photon density in terms of time through the laser resonator is expressed by

$$\frac{d\phi}{dt} = c\phi\sigma N - \frac{\phi}{\tau_c} \tag{2.69}$$

where $\tau_c = l/\gamma c_0$ is the photons decay time in the resonator, l is the effective resonator length, $\gamma = (\delta + \gamma_1 + \gamma_2)/2$ is the total losses per pass in the resonator, δ is the internal cavity losses and $\gamma_{1,2} = -lnR_{1,2}$ where R_1 and R_2 are the reflectivities of the laser cavity mirrors.

Under steady-state conditions, the population inversion at above threshold ($\phi \rangle 0$) is given by:

$$N = \frac{(f_a + f_b)W_b N_g \tau + N^0}{1 + (f_a + f_b)c\sigma\phi\tau} \tag{2.70}$$

By setting $\phi = 0$ a formula for the unsaturated gain in the crystal is obtained:

$$g_0 = \sigma N_c = (f_a + f_b)W_b N_g \tau\sigma - \sigma f_a N_g \tag{2.71}$$

In this situation, the unsaturated gain in the active medium is extremely related to the re-absorption losses owing to the lower laser level population. By supposing proportion between the pump rate, W_p, and the input power of flashlamp, P_{in}, and using g_0 in the formula for the threshold condition:

$$2g_0 l = 2\theta P_{in}(th) \tag{2.72}$$

where $\theta = \eta_p\big/\big[I_s A(f_a + f_b)\big]$ is the effective pumping coefficient, $I_s = h\nu/\sigma\tau(f_a + f_b)$ is the saturation intensity for the laser transition at frequency ν, and η_p and l are the pumping coefficient related to medium characteristics, and length of the laser medium respectively.

At threshold condition, gain coefficient can be expressed as

$$g_0 = \left[\frac{1}{2l}\ln\frac{1}{Rr} + \alpha\right] \tag{2.73}$$

Therefore P_{th} can be written as

$$P_{th} = \frac{h\nu V}{\eta_p \sigma\tau}\left[\frac{1}{2l}\ln\frac{1}{Rr} + \alpha\right] \tag{2.74}$$

where R, r, τ, α and V are the reflectivity of both mirrors at corresponding transition frequency of ν, the fluorescent lifetime at the upper level, loss at the corresponding emission frequency and the laser rod volume respectively.

$$2g_0 l = (2\gamma_i + 2l\sigma f_a N_g) - \ln R_1 \tag{2.75}$$

We obtain at $P_{in} = P_{th}$:

$$P_{th} = \frac{\delta - \ln R_1}{2\theta} \tag{2.76}$$

where $\delta = 2\gamma_i + 2l\sigma f_a N_g$ is the internal round-trip cavity losses, comprised of losses owing to re-absorption from the lower laser level and other items indicated as γ_i. Eq. (2.76) shows the threshold equation for quasi three level system is similar to the four-level system but the material parameters is different.

The threshold and slope efficiency are important parameters of laser operation. The output power of laser emerges while threshold input power is reached and at above threshold Eq. (2.76) is modified by Eq. (2.59) to give the power input as

$$P_{in} = \frac{\delta - \ln R_1}{2\theta}\left(1 + \frac{I}{I_s}\right) = P_{in}(th)\left(1 + \frac{I}{I_s}\right) \qquad (2.77)$$

The output power above threshold is expressed by the equation

$$P_{out} = \eta_s\left[P_{in} - P_{in}(th)\right] \qquad (2.78)$$

where η_s is the slope efficiency.

2.7. TEMPERATURE DEPENDENCE OF LONG PULSE LASERS

The performance of free running or long pulse of Nd:YAG laser, can be quantified based on the output E_{out} laser which expressed:

$$E_{out} = \eta_s\left(E_{in} - E_{th}(T)\right) \qquad (2.79)$$

where $E_{th}(T)$ is the threshold energy, η_s is the slope efficiency, and E_{in} is the input energy. The stimulated emission cross section of Nd:YAG laser rod is dependent on operating temperature of the water cooling system. Therefore the threshold energy $E_{th}(T)$ or the minimum pumped energy required to initiate lasing also depends on temperature. The threshold energy is expressed by

$$E_{th}(T) = \frac{Vh\nu_p}{2\sigma(T)\eta_{pe}}(-lnR + L)$$
(2.80)

where V is the laser volume, R is the reflectivity of the resonator output coupler, L is intracavity nonproductive losses, σ is stimulated emission cross-section, η_{pe} is the efficiency with which pump photon at a frequency ν_p are absorbed into the upper laser level. Differentiating (2.70) and (2.80) with respect to temperature leads to

$$\frac{dE_{out}}{dT} = \eta_s\left(-\frac{dE_{th}}{dT}\right)$$
(2.81)

The Equation (2.81) shows that the laser output energy proportional to the slope efficiency and the change in threshold energy but the laser output effectively decreases as temperature increases.

$$\frac{1}{E_{th}}\frac{dE_{th}}{dT} = \left(-\frac{1}{\sigma}\frac{d\sigma}{dT}\right)$$
(2.82)

The Equation (2.82) explained that the threshold energy E_{th} is increased as the temperature increases if the change in cross-section with respect to temperature $d\sigma/dT$ is negative. In this case, the input energy remains constant at each temperature. The slope efficiency η_s is determined by temperature-independent properties of the resonator, pump geometry, and absorption characteristics of the laser medium.

2.8. THEORETICAL ANALYSIS OF OSCILLATION CONDITION OF SIMULTANEOUS DUAL-WAVELENGTH

Rare earth elements possess many energy levels, which will be split into a number of Stark sublevels due to the action of crystal field. The transitions between the Stark sublevels of different manifolds can radiate lights at various frequencies. Each transition line has different stimulated emission cross-section, fluorescent lifetime, and fluorescent quantum efficiency, therefore only one oscillation can oscillate in the same crystal. But under special conditions, more than one wavelength can be achieved simultaneously.

Laser systems only one fundamental wavelength operation can normally oscillate if no special measure is taken. The emission lines with weak gain are usually depressed by the line with strong gain because of the gain competition between the laser emission lines, and this competition generally results in only the laser line with the strongest gain being generated. To meet the requirement of simultaneously multiple laser line oscillation, the special design of the laser oscillation cavity is necessary to control and to reduce the gain competition among the multiple wavelength lines of the gain medium.

If suppose n laser transition in a laser medium, the condition must be satisfied for each oscillation according to the Eq. (2.45):

$$R_i r_i \exp 2(g_i - \alpha_i)l = 1 \quad i=1, 2, ..., n \tag{2.83}$$

where R_i, r_i, g_i and α_i are the reflectivity of the rare mirror, the reflectivity of output coupler, the gain and loss at the corresponding emission wavelength respectively. l is the length of the laser medium.

Eq. (2.76) gives the pump power input to the lamp at the threshold for each transition line

$$P_{thi} = \frac{h v_i V}{\eta_i \sigma_i \tau_i}[\alpha_i + \frac{1}{2L}(\frac{1}{R_i r_i})] \tag{2.84}$$

Here σ_i is the stimulated emission cross section for the corresponding transition, τ_i is the fluorescent lifetime at the upper level, V is the volume of rod, η_i is the pump efficiency.

In order to achieve dual frequency oscillation in a cavity, both transitions must possess the same threshold power. For several neodymium embedded crystals such as Nd:YAG, Nd: BEL, Nd: YAP, and Nd: YLF crystals, the efficiency, and losses of the crystal are almost equal for all of the laser transitions in each crystal. In addition, if transition lines originate from the same upper laser level, the fluorescence lifetimes of the upper laser level are the same for all the wavelength operation. Therefore we have

$$\ln(\frac{1}{R_2 r_2}) = 2\alpha L[\frac{\sigma_2 V_1}{\sigma_1 V_2}-1]+\frac{\sigma_2 V_1}{\sigma_1 V_2}\ln(\frac{1}{R_1 r_1}) \tag{2.85}$$

According to Eq. (2.85), with different reflectivity values of output mirror for the emission line at the $1=1064$ nm, the corresponding reflectivity values for the emission at $2=946$ nm is calculated. The results are shown in Figure 2.5. The basic parameters used in the calculation are $\sigma_2 = 4.6 \times 10^{-19}$ cm^2, $\sigma_1 = 4 \times 10^{-20}$ cm^2, L = 7 cm and $\alpha = 0.0026$ cm^{-1}.

Furthermore, when the reflectivity values of output mirror at the 946 nm laser transition are in the range of 70%-99%, the reflectivity of output mirror at the 1064 nm transition must be in the range of 3.6%-58% for Nd:YAG crystal.

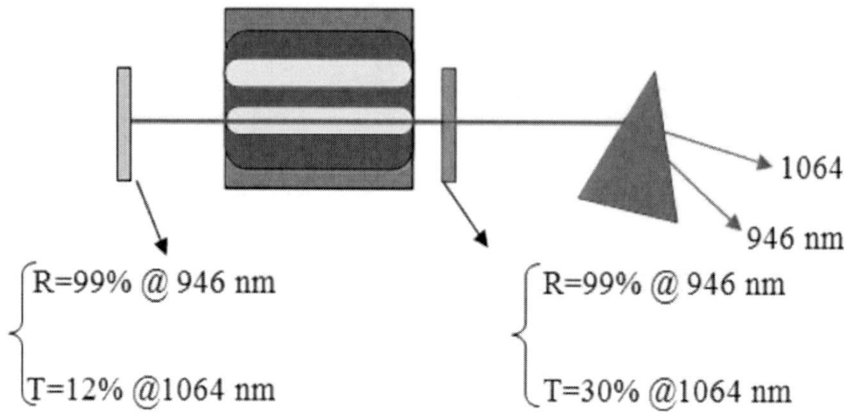

Figure 2.5. Experimental setup of dual wavelength generation at 1064 and 946 nm.

2.9. OPTICAL PROPERTIES OF ND:YAG CRYSTAL

Synthetic garnets such as gadolinium scandium aluminum garnet [$Gd_3Sc_2Al_3O_{12}$ (GSGG)], yttrium aluminum garnet [$Y_3Al_5O_{12}$ (YAG)], and gadolinium gallium garnet [$Gd_3Ga_5O_{12}$ (GGG)] have many desirable features as laser host materials. They are, hard, stable optically isotropic, and have good thermal conductivities, which causes lasers mediums operate at high power levels. At present, a variety of laser materials has been developed, among which the most standard host is the yttrium aluminum garnet (YAG). Besides, the cubic form of YAG has a narrow emission linewidth, which permits low threshold and high gain for laser operation.

The host crystal YAG can be doped with rare-earth (RE) elements to produce laser gain and amplification systems, such as Tm: YAG, Er: YAG and Yb: YAG. REs are used as a dopant due to their strong absorption bands, low non-radiative decay rates, and sharp transition lines in the spectra. Owing to remarkable laser properties of Nd^{3+} doped YAG crystal, such as high mechanical strength, thermal conductivity, optical transparency over a wide spectral region, adequate fluorescence lifetime for storage energy and high stimulated emission cross-section, it has been utilized for a long time in the solid-state laser industry.

The electronic structure of the Nd^{3+} ion is: $1s^2\,2s^2\,2p^6\,3s^2\,3p^6\,3d^{10}\,4s^2\,4p^6\,4d^{10}\,4f^3\,5s^2\,5p^6$. The structure of the energy level free Nd^{3+} ion is determined by the electrons Coulomb interactions with the nucleus and with each other, and also with the spin and orbital moments of the electrons. Hence, it belongs to a $4I_J$ atomic state which is occupied by three electrons of the Nd^{3+} ion and being on the 4f shell, which have a total spin moment S of 3/2 and orbital moment $L = 6$. Because of the transition inside the 4f-electronic shell is screened by 8 outer electrons ($5s^2$ and $5p^6$), the spectroscopic properties of the Nd:YAG is determined by the basic properties of a Nd^{3+} ion and weakly influenced by the YAG crystalline matrix.

The symbol characterizing each level is in form $^{2S+1}L_J$, where S is the total spin quantum number, J is the total angular momentum quantum number and L is the orbital quantum number. The allowed values of L, namely $L= 0, 1, 2, 3, 4, 5, 6\ldots$ are expressed by the capital letters $S, P, D, F, G, H, I, \ldots$, respectively. Thus the $^4I_{9/2}$ ground level corresponds to a state in which $2S+1 = 4$ (so $S = 3/2$), $L= 6$ and $J = 9/2$. Each $^{2S+1}L_J$ is split into $(2J+1)/2$ degenerate sublevels. Thus the $^4I_{9/2}$, $^4F_{3/2}$, and $^4I_{11/2}$ are split into 5, 2 and 6 sublevels, respectively.

2.10. ENERGY LEVELS OF ND:YAG CRYSTAL

The energy level diagram of the Nd:YAG crystal with various spin-orbits Stark splits by the crystal field at 300 K is illustrated in Figure 2.6. Pumping transfers the Nd ions from ground state $^4I_{9/2}$ to several relatively narrow absorption bands. Typically the pump bands of the Nd:YAG occurred at two main levels, $^4F_{5/2}$ and $^4F_{7/2}$ on the wavelength of 810 nm and 750 nm, respectively (Koechner 2006; Svelto 2010). These band then rapidly relax non-radiatively to the $^4F_{3/2}$ level and followed into lower-lying $^4I_{13/2}$, $^4I_{11/2}$, and $^4I_{9/2}$ manifolds. The $^4F_{3/2}$ level is metastable because the next lower level is separated from it by 4698 cm^{-1}. The specific absorption and emission wavelengths depend on the crystal field, which influences the splitting within a manifold and between different manifolds.

In RE elements *4f* subshell is incomplete filled and shielded by the completely filled outer *5s* and *5p* subshells. Therefore, narrow absorption and emission lines independent of the host material emerge when REs embedded into crystals or glasses [72]. As shown in Figure 2.6, Coulomb interaction produces energetic dividing between the ^4F and ^4I levels of the *4f* electrons. However spin-orbital interaction of the electrons causes dividing between the *4I* (and also between *4F*) intermanifolds levels. In addition, electrical crystal field divides each manifold into Stark sublevels [73].

Nd^{3+} doped solid-state lasers have been demonstrating on four transitions from the $^4F_{3/2}$ manifold in the wavelength regions 0.93 (quasi-three-level system), 1.05, 1.32, and 1.8 µm (four-level system) corresponding to lower manifolds of $^4I_{9/2}$, $^4I_{11/2}$, $^4I_{13/2}$, and $^4I_{15/2}$, respectively.

The majority of these lasers operate on four level system either the $^4F_{3/2}-$ $I_{11/2}$ or $^4F_{3/2}-^4I_{13/2}$ transitions. One difficulty with the $^4F_{3/2}-^4I_{9/2}$ transition is because in the quasi-three-level system the transitions are disordered by considerable re-absorption loss due to the thermal population at a lower level, meanwhile, in four-level systems, the re-absorption loss at transitions $^4F_{3/2}-^4I_{11/2}$ and $^4F_{3/2}-$ $^4I_{13/2}$ are considered neglected. In addition, the emission cross-section of transition $^4F_{3/2}-$ $^4I_{9/2}$ is much smaller than those of transitions $^4F_{3/2}-$ $^4I_{11/2}$ and $^4F_{3/2}-$ $^4I_{13/2}$ and thus leads to lasing on the higher gain transitions. Transition lines emitted from a Nd^{3+}: YAG crystal was shown in Figure 2.8.

Nd:YAG can create several transition lines from upper laser level $^4F_{3/2}$ to lower laser levels $^4I_{15/2}$, $^4I_{13/2}$, $^4I_{11/2}$ (four-level transition) and $^4I_{9/2}$ (quasi-three-level transition) manifolds respectively. The branching ratio (β) for these transitions has been found to be $^4F_{3/2}\rightarrow^4F_{15/2} = 0.01$, $^4F_{3/2}\rightarrow^4F_{13/2} = 0.14$, $^4F_{3/2}\rightarrow^4F_{11/2} = 0.60$, $^4F_{3/2}\rightarrow^4F_{9/2} = 0.25$.

The strongest transition line at room temperature is the four-level transition line at 1064 nm and threshold power of quasi-three-level laser transitions are higher than for four level laser transition lines. Small stimulated emission cross section and re-absorption loss are the main problems to achieve the quasi-three-level emissions.

The energy range between the $^4I_{11/2}$ state and the ground state $^4I_{9/2}$ is approximately 2000 cm^{-1} which is an order of magnitude higher than the value kT, and this result in the four-level nature of lasing on this transition (Antisiferov, 2005).

In Nd doped crystals the efficient quasi-three-level lasing transitions are in the range of 900-950 nm. The strongest lines of quasi-three-level transition in Nd:YAG is 938.5 and 946 nm. However, the transition at 946 nm has a very small stimulated emission cross-section and the terminal laser level located at 852 cm^{-1} above the ground level. There is a thermal population at room temperature which causes significant reabsorption at the laser wavelength (Fan and Byer, 1987; Koechner, 2006). This increases the internal power loss in the laser medium which in turn contributes to the extra heat in the laser crystal.

Therefore, to achieve stable operation at quasi-three-level lines not only need a low-loss resonator design by means of precise mirrors and using the cooling system but also need effective suppression of other stronger laser transitions, such as 1064 and 1319 nm.

Figure 2.6 shows energy levels, some absorption and emission lines of Nd:YAG crystal. The lines of 938.5 and 946 nm emissions are from sublevels R_1 and R_2 of the two crystal-field components of the upper laser level $^4F_{3/2}$ to the lower laser level (Z_5) of the five crystal-field components of the ground level $^4I_{9/2}$. The red arrows indicate absorption lines and their probabilities, blue arrows show quasi-three-level system at 938.5 and 946 nm and four-level transition lines and dashed lines show non-radiative decays respectively.

The pump bands are $^4F_{3/2}$ level and another level where located above the upper laser level. The pump band of manifold $^4F_{5/2}$ is responsible for absorption peak around 808 nm which is the basic absorption wavelength for laser diode pumping. The main absorption band of Nd:YAG is claimed at 0.525 to 0.585 μm, 0.73 to 0.75 μm and 0.79 to 0.9 μm as illustrated in Figure 2.7. Since flashlamps produce a wide range of radiation, The absorption wavelengths are important for flashlamp-pumped Nd:YAG lasers.

Generally, an optically pumped Nd:YAG crystal exhibits several another absorption peaks at 300 K, mostly at $^4F_{3/2}$ (880 nm), $^4G_{7/2} + ^2G_{5/2}$ (580 nm) and $^2K_{13/2} + ^4G_{7/2} + ^4G_{9/2}$) (520 nm) as shown in Figure 2.6. However, The multiplets above the $^4F_{7/2}$ and $^4S_{3/2}$ states are practically not involved in the energy balance. In the case of $Y_3Al_5O_{12}$ crystal excited by a xenon flashlamp, estimates show that they account for only 10% of the total absorbed excitation energy (Kaminskii; 2006). The absorption coefficient of the pumping light is determined by the effective cross-section of the transition (Zagumenyi et al., 2004; Svelto, 2010; Koechner, 2006).

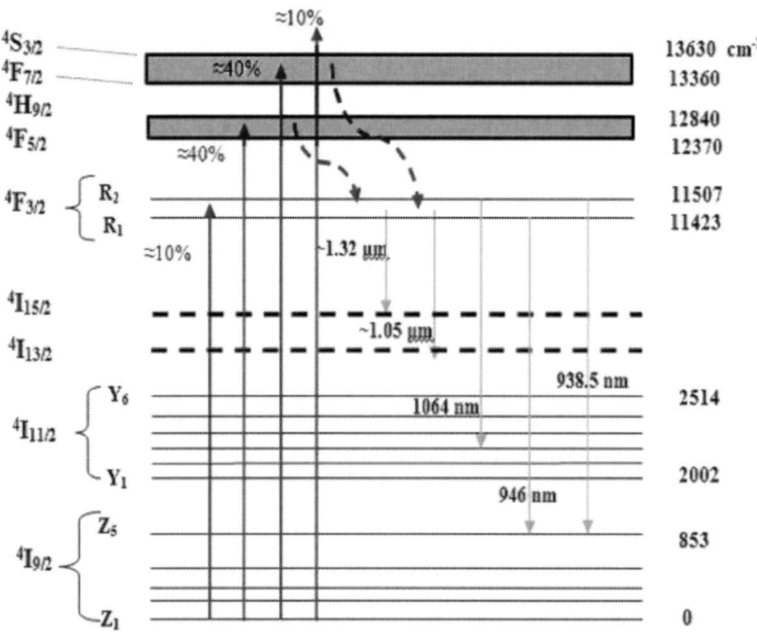

Figure 2.6. Energy levels of Nd:YAG.

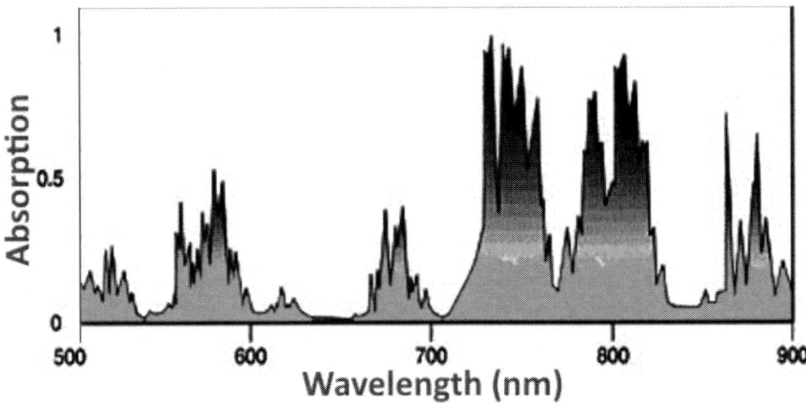

Figure 2.7. Absorption spectrum of Nd:YAG.

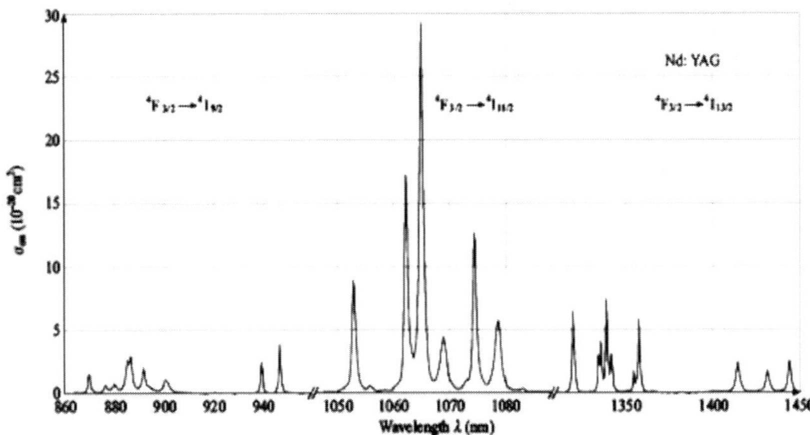

Figure 2.8. Emission spectrum of the Nd:YAG crystal at 300K [73].

REFERENCES

[1]   Wallace, S. E. W. a. H. (1969). "Oscillation and doubling of the 0.946-µ line in Nd:YAG", *Applied Physics Letter*, *15*, 111-112.

[2]   Birnbaum, M., Tucker, A. W. & Pomphrey, P. J. (1972). "New Nd:YAG Laser Transitions 4F3/2→4I9/2", *IEEE J. of Quantum electron*, *8*, 501-501.

[3]   Fan, T. Y. & Byer, R. L. (1987). "Continuous wave operation of a room temperature, diode-laser pumped, 946 nm Nd:YAG laser", *Optics Letter*, *12*, 809-811.

[4]   Risk, W. P. & Length, W. (1987). "Room temperature, continuous wave, 946nm Nd:YAG laser, pumped by laser-diode arrays and intracavity frequency doubling to 473 nm", *Optics Letter*, *12*(993-995).

[5]   Risk, W. P. (1988). "Modeling of longitudinally pumped solid-state lasers exhibiting reabsorption losses", *J. Opt. Soc. Am. B*, *5*, 1412-1423.

[6]   Zhou, R., Cai, Z., Wen, W., Ding, X., Wang, P. & Yao, J. (2005). "High-power continuous-wave Nd:YAG laser at 946 nm and intracavity frequency-doubling with a compact three-element cavity", *Optics Communications*, *225*, 304-308.

[7]   Dixon, G. J., Zhang, Z. M., Chang, R. & Djeu, N. (1988). "Efficient blue emission from an intracavity-doubled 946-nm Nd:YAG laser", *Optics Letter*, *13*, 137-139.

[8]   Li, P. X., Li, D. H., Zhang, Z. G. & Zhang, S. W. (2003). "Diode-pumped high power cw blue laser at 473 nm with a compact three-element cavity", *Optics Communications*, *215*, 159-162.

[9]   Li, P. X., Li, D. H., Zhang, Z. G. & Zhang, S. W. (2003). "Diode pumped compact cw frequency doubled Nd:YAG laser in the watt range at 473 nm", *Chinese Physics Letters*, *20*, 1064-1066.

[10]   Rui, Z., Young, Z. S., Qiang, C. Z., Qi, W., Xin, D., Peng, W., Li, D. & Quan, Y. (2005). "High Power Diode-End-Pumped Nd:YAG 946-nm Laser and Its Eficient Frequency Doubling", *Chinese Physics Letters*, *22*, 1413-1415.

[11]   Ling, Z., Yong, L. C., Hua, F. B., Yi, W. Z., Hua, L. D., Ming, F. U. & Guo, Z. Z. (2005). "Diode-Pumped Passive Q-Switched 946-nm Nd:YAG Laser with 2.1-W Average Output Power", *Chinese Physics Letters*, *22*, 1420-1422.

[12] Chen, Y., Peng, H., Hou, W., Peng, Q., Geng, A., Guo, L., Cui, D. & Xu, Z. (2006). "3.8 W of cw blue light generated by intracavity frequency doubling of a 946-nm Nd:YAG laser with LBO", *Appied Physics B*, *83*, 241-243.

[13] Bethea, C. G. (1973). "Megawatt Power at 1.318 μm in Nd3+:YAG and Simultaneous Oscillation at Both 1.06 and 1.318 μm", *IEEE J. OF Quantum. Electron*, *9*, 254-254.

[14] Vollmar, W., Knights, M. G., Rines, G. A., McCarthy, I. C. & Chicklis, E. P. (1983). "Five-color Nd:YLF laser", *Digest of Conference on Lasers and Electro Optics*, 188-188.

[15] Shi, W. Q., Kurtz, R., Machan, J., Bass, M., Birnbaum, M. & Kokta, M. (1987). "Simultaneous, multiple wavelength lasing of (Er, Nd):Y3Al5O12", *Applied Physics Letter*, *51*, 1218-1220.

[16] Machan, J., Kurtz, R., Bass, M., Birnbaum, M. & Kokta, M. (1987). "Simultaneous, multiple wavelength lasing of (Ho, Nd):Y3Al5O12", *Applied Physics Letter*, *51*, 1313-1315.

[17] Nadtocheev, O. E. V. E. a. N. (1989). "Two-wave emission from a cw solid-state YAG:Nd3+ laser", *Soviet. J. of Quantum Electronics*, *19*, 444.

[18] Shen, H. Y., Zeng, R. R., Zhou, Y. P., Yu, G. F., Huang, C. H., Zeng, Z. D., Zhang, W. J. & Ye, Q. J. (1990). "Comparison of simultaneous multiple wavelength lasing in various neodymium host crystals at transition from 4F3/2–4I11/2 and 4F3/2–4I13/2", *Applied Physics Letter*, *56*, 1937-1938.

[19] Shen, H. Y., Zeng, R. R., Zhou, Y. P., Yu, G. F., Huang, C. H., Zeng, Z. D., Zhang, W. J. & Ye, Q. J. (1991). "Simultaneous multiple wavelength laser action in various neodymium host crystals", *IEEE J. of Quantum Electron*, *27*, 2315-2318.

[20] Wang, C. Q., Chow, Y. T., Yuan, D. R., Xu, D., Zhang, G. H., Liu, M. G., Lu, J. R., Shao, Z. S. & Jiang, M. H. (1999). "Cw dual-wavelength Nd:YAG laser at 946 and 938.5 nm and intracavity nonlinear frequency conversion with a CMTC crystal", *Optics Communications*, *165*, 231-235.

[21] Li, P., Li, D., Li, C. & Zhang, Z. (2004). "Simultaneous dual-wavelength continuous wave laser operation at 1.06 lm and 946 nm in Nd:YAG and their frequency doubling", *Optics Communications*, *235*, 169-174.

[22] Lee, Y. P. H. C. a. K., (2008). "Simultaneous dual-wavelength oscillation at 1357 nm and 1444 nm in a Kr-flashlamp pumped Nd:YAG laser", *Optics Communications*, 281 4455-4458.

[23] Zhu, H., Zhang, G., Huang, C. H., Wei, Y., Huang, L. X., Li, A. H. & Chen, Q. (2008). "1318.8/1338.2 nm simultaneous dual-wavelength Q-switched Nd:YAG laser", *Applied Physics B*, *90*, 451-454.

[24] Huan, L., Gang, X. D. & Quan, Y. (2009). "Simultaneous all-solid-state multi-wavelength Lasers: a promising pump source for generating highly coherent terahertz waves", *Chinese Physics B*, *18*, 1077-1084.

[25] Li, C. Y., Bo, Y., Xu, J. L., Tian, C. Y., Peng, Q. J., Cui, D. F. & Xu, Z. Y. (2011). "Simultaneous dual-wavelength oscillation at 1116 and 1123 nm of Nd:YAG laser", *Optics Communications*, *284*, 4574-4576.

[26] Hongyuan, S. (1990). "Oscilation condition of simultaneous multiple wavelength lasing", *Chinese Physics Letters*, 7, 174-176.

[27] Zhou, R., Li, E. B., Zhang, G., Ding, X., Cai, Z. Q., Wen, W. Q., Wang, P. & Yao, J. Q. (2006). "Simultaneous dual-wavelength CW operation using 4F3/2–4I13/2 transitions in Nd:YVO4 crystal", *Optics Communications*, *260*, 641-644.

[28] Wu, B., Jiang, P. P., Yang, D. Z., Chen, T., Kong, J. & Shen, Y. H. (2009). "Compact dual wavelength Nd:GdVO4 laser working at 1063 and 1065 nm", *Optics Express*, *17*, 6004-6009.

[29] Yu, H. H., Zhang, H. J., Wang, Z. P., Wang, J. Y., Yu, Y. G., Shi, Z. B., Zhang, X. Y. & Jiang, M. H. (2009). "High power dual wavelength laser with with disordered Nd:CNGG crystals", *Optics Letter*, *34*, 151-153.

[30] Maestre, H., Torregrosa, A. J., Pousa, C. R., Rico, M. L. & Capmany, J. (2008). "Dual-wavelength green laser with a 4.5 THz frequency difference based on self-frequency- doubling in Nd3+-doped aperiodically poled lithium niobate", *optics letter, 33,* 1008-1010.

[31] Yoshioka, H., Nakamura, S., Ogawa, T. & Wada, S. (2010). "Dual-wavelength mode-locked Yb:YAG ceramic laser in single cavity", *Optics Express, 18,* 1479-1486.

[32] Janousek, J., Lichtenberg, P. T., Mortensen, J. L. & Buchhave, P. (2006). "Investigation of passively synchronized dual-wavelength Q-switched lasers based on V:YAG saturable absorber", *Optics Communications, 265,* 277-282.

[33] Yu, H. H., Zhang, H. J., Wang, Z. P., Wang, J. Y., Yu, Y. G., Zhang, X. Y., Lan, R. J. & Jiang, M. H. (2009). "Dual-wavelength neodymium-doped yttrium aluminum garnet laser with chromium-doped yttrium aluminum garnet as frequency selector", *Applied Physics Letter,* 94 041126.

[34] Shen, H. Y. & Su, H. (1999). "Operating conditions of continuous wave simultaneous dual wavelength laser in neodymium host crystals", *Applied physics, 86,* 6647-6651.

[35] Guo, L., Lan, R. J., Liu, H., Yu, H. H., Zhang, H. J., Wang, J. Y., Hu, D. W., Zhang, S. D., Chen L. J., Zhao, Y. G., Xu, X. G. & Wang, Z. P. (2010). "1319 nm and 1338 nm dual-wavelength operation of LD end pumped Nd:YAG ceramic laser", *Optics Express, 18,* 9098-9106.

[36] Xue, L. P., Hua, L. D., Yong, L. C. & Guo, Z. Z. (2004). "Oscillation conditions of cw simultaneous dual-wavelength Nd:YAG laser for transitions 4F3/2 → 4I9/2 and 4F3/2 →4I11/2", *chinese physics, 13,* 1689-1693.

[37] Bjurshagen, S., Evekull, D. & Koch, R. (2002). "Generation of blue light at 469 nm by efficient frequency doubling of diode-pumped Nd:YAG laser", *Electronics Letters, 38,* 324-325.

[38] Pavel, N. (2010). "Simultaneous Dual-Wavelength Emission at 0.90 and 1.06 µm in Nd-doped Laser Crystals", *laser physics*, *20*, 215–221.

[39] McCumber, D. & Sturge, M. (1963). "Linewidth and temperature shift of the R lines in Ruby", *Applied Physics*, *34*, 1682-1984.

[40] Vink, A. P. & Meijerink, A. (2000). "Electron-phonon coupling of Cr3+ in YAG and YGG", *Journal of Luminescence*, *87*, 601-604.

[41] Macfarlane, R. M. (2000). "Direct process thermal line broadening in Tm:YAG", *Journal of Luminescence*, *85*, 181-186.

[42] Sardar, D. K., Yow, R. M. & Salinas, F. S. (2001). "Stark components of lower lying manifolds and phonon effects on sharp spectral lines for inter-Stark transitions of Nd3+ in LLGG crystal host", *Optical Materials*, 18 301-308.

[43] Sardar, D. K. & Stubblefield, S. C. (1998). "Temperature dependence of linewidths, positions, and line shifts of spectral transitions of trivalent neodymium ions in barium magnesium yttrium germinate laser host", *Applied physics*, *83*, 1195-1199.

[44] Johhson, S. A., Freie, H. G., Schawlowa, A. L. & Yen W. M. (1967). "Thermal shifts in the energy levels of LaF3:Nd3+", *Journal of the Optical Society of America*, *57*, 734-737.

[45] Sardar, D. K., Yow, R. M. & Sayka, A. (2001). "Crystal-Field Splittings and Phonon Effects on a Sharp Emission Line within a Manifold of Pr3+ in Ca5(PO4)3F Laser Host", *Phys. Stat. Sol. (b)*, *223*, 691-700.

[46] Kushida, T. (1989). "Linewidths and thermal shifts of spectral lines in neodymium-doped yttrium aluminum garnet and calcium fluorophosphates", *Physical Review*, *185*, 500-508.

[47] Xing, J. C. S. Z. a. B. (1988). "Thermal shifts of the spectral lines in the 4F3/2 to 4I9/2", *IEEE journal of quantum electronics*, *24*, 1829-1832.

[48] Sardar, R. M. D. K. a. Y. (1998). "Optical characterization of inter-Stark energy levels and effects of temperature on sharp emission lines of Nd3+ in CaZn2Y2Ge3O12", *Optical Materials*, *10*, 191-199.

[49] Sardar, D. K., Yow, R. M. & Salinas, F. S. (2001). "Stark components of lower-lying manifolds and phonon effects of sharp spectral lines for inter-Stark transitions of Nd3+ in LLGG crystal host", *Optical Materials*, *18*, 301-308.

[50] Mao, Y., Huang, M. & Wang, C. (2004). "Temperature effect on emission lines and fluorescence lifetime of the 4F3/2 state of Nd:YVO4", *Chinese Optics Letters*, *2*, 102-105.

[51] Singh, S., Smith, R. G. & Van, L. G. (1974). "Stimulated emission cross section and fluorescent quantum efficiency Of Nd3+ in yttrium aluminium garnet at room temperature", *Physics Review B*, *10*, 2566-2572.

[52] Kaminskii, A. A. (1990). *Laser Crystals* (second edition, Springer series in optical science).

[53] Kushida, T., Marcos, H. M. & Geusic, J. E. (1968). "Laser transition cross section and fluorescence branching ratio for Nd3+ for yttrium aluminum garnet", *Physics Review*, *167*, 289-291.

[54] Krupke, W. F., Shin, M. D., Marion, J. E., Carid, J. A. & Stokowski, S. E. (1986). "Spectroscopic, optical, and thermomechanical properties of neodymium- and chromium-doped gadolinium scandium gallium garnet", *J. Opt. Soc. Am. B.*, *3*, 102-114.

[55] Powel, R. C. (1998). *Physics of solid laser materials* (AIP Press/Springer, New York).

[56] Rapaport, A., Zhao, S., Xiao, G., Howard, A. & Bass, M. (2002). "Temperature dependence of the 1.06-μm stimulated emission cross section of neodymium in YAG and in GSGG", *Applied physics*, *41*, 5052-7057.

[57] Zhaoa, S., Rapaport, A., Dong, J., Chen, B., Deng, P. & Bass, M. (2006). "Temperature dependence of the 1.064-mm stimulated emission cross-section of Cr:Nd:YAG crystal", *Optics & Laser Technology*, *38*, 645–648.

[58] Bass, M., Weichman, L. S., Vigil, S. & Brickeen, B. K. (2003). "The Temperature dependence of Nd3+ doped solid-state lasers", *IEEE J. of Quantum Electronics*, *39*, 741-748.

[59] Sardar, D. K., Yow, R. M., Gruber, J. B., Allik, T. H. & Zandi, B. (2006). "Stark components of lower-lying manifolds and emission cross-sections of intermanifold and inter-stark transitions of Nd+3 (4f3) in polycrystalline ceramic garnet Y3Al5O12", *Journal of Luminescence*, *116*, 145-150.

[60] Kumar, G. A., Lu, J., Kaminskii, A. A., Ueda, K. I., Yagi, H., Yanagitani, T. & Unnikrishnan, N. V. (2004). "Spectroscopic and Stimulated Emission Characteristics of Nd3+ in Transparent YAG Ceramics", *IEEE J. of Quantum electronics*, *40*, 747-758.

[61] Krupke, W. F., Shin, M. D., Marion, J. E., Carid, J. A. & Stokowski, S. E. (1986). "Spectroscopic, optical, and thermomechanical properties of neodymium- and chromium-doped gadolinium scandium gallium garnet", *J. Opt. Soc. Am. B*, *3*, 102-114.

[62] Svelto, O. (2010). *Principles of laser.* (5th ed.). New York: Springer.

[63] Luo, Z., Huang, Y. & Chen, X. (2007). *"Spectroscopy of solid-state laser and luminescent materials"*, Chinese Academy of Sciences.

[64] Powell, R. C. (1998). *"Physics of solid-state laser materials"*, Springer-Verlag.

[65] Bartolo, B. D. (2010). *Optical interaction in solids*, 2nd edition (Boston College, USA).

[66] Chen, B. D. X. a. B. (1993). "Phonon effects on sharp luminescence lines of Nd3+ in Gd3Sc2Ga3O12 garnet (GSGG)", *Journal of Luminescence*, *54*, 309-318.

[67] Csele, M. (2004). *Fundamentals of light sources and lasers.* (1st ed.). New Jersey: John Wiley & Sons, Inc.

[68] Li, F. Q., Zhang, X. F., Zong, N., Yang, J., Peng, Q. J., Cui, D. F & Xu, Z. Y. (2009). "High-Efficiency High-Power Nd:YAG Laser under 885nm Laser Diode Pumping", *Chinese Physics Letters, 26,* 114206-3.

[69] Quimby, R. (2006). *Photonics and Laser: An Introduction.* (1st ed.). New Jersey: John Wiley & Sons.

[70] Foot, C. J. (2005). *Atomic Physics.* (1st ed.). USA: Oxford University Press.

[71] Fan, T. Y. & Byer, R. L. (1987). "Modeling and cw operation of a quasi three level 946 nm Nd:YAG Laser", *IEEE J. of Quantum Electonics, 23,* 605-612.

[72] Verlinden, N. (2008). *"The Excited State Absorption Cross Section of Neodymium-doped Silica Glass Fiber in the 1200-1500 nm Wavelength Range"*, Master of sciences thesis, Worcester Polytechnic Institute.

[73] Trager, F. (2007). *Springer Handbook of laser and optics.* (1st ed.) New York: Springer.

Chapter 3

RESEARCH METHODOLOGY OF ND^{3+}: YAG LASER PUMPED BY FLASHLAMP

ABSTRACT

In order to form an operational and efficient laser system, several components such as resonator, power supply unit and cooling system are joint together. Using this homemade laser system this work is divided into three main tasks. First, the emission spectra, second, the shift and broadening of emission lines in Nd^{3+}:YAG laser crystal and third, the temperature dependence of quasi-three-level transition. The laser system comprised of a homemade flashlamp power supply. A Nd:YAG laser rod manufactured from Calstec is employed as an active medium. For most lasers, cavity mirrors are vital to increasing the circulating power within the cavity where gain overcomes losses. A linear flashlamp filled xenon gas at a pressure of 450 Torr is placed parallel to the laser rod in a ceramic reflector for preparation of side pumping technique.

Keywords: power supply, cooling system, Nd^{3+}:YAG laser crystal, flashlamp

3.1. DESIGN OF COMPONENTS OF THE Nd^{3+}:YAG LASER PUMPED BY FLASHLAMP

Several components such as resonator, power supply unit and cooling system are joint together in order to form an operational and efficient laser system. In the following sections, the major components of the homemade Nd^{3+}:YAG laser pumped by flashlamp which was designed and constructed by previous researchers will be discussed.

By using this homemade laser system the work is divided into three main tasks. Firstly, the emission spectra of the flashlamp and the Nd^{3+}:YAG laser rod pumped by Xenon flashlamp were investigated. Particular attention was paid to spectroscopy characteristics of quasi-three-level and four level laser transitions of Nd^{3+}:YAG crystal.

Secondly shift and broadening of emission lines in Nd^{3+}:YAG laser crystal influenced by input energy and temperature as well as effects of temperature and input energy on the quasi-three-level and four level emissions cross section of Nd^{3+}:YAG were analyzed. The temperature dependence of quasi-three-level transition for long pulse Nd^{3+}:YAG laser was investigated.

3.2. ND:YAG LASER SYSTEM

The main components in homemade Nd:YAG laser system comprised of flashlamp driver or power supply to energize the flashlamp and the laser head. The detail each of these components will be discussed in the following sections.

3.2.1. Power Supply Triggered by a Simmer Mode Technique

The laser system comprised of a homemade flashlamp power supply such as shown in Figure 3.1. The driver has been designed by the previous researcher. The power supply is triggered by a simmer mode technique. Five major parts of the power supply are including of capacitor charger power supply (CCPS), pulse forming network (PFN), ignition circuit, simmer circuit and control unit. The voltage output of the power supply was variable from 0 - 1000 V. The capacitor (150 μF) was charged to have the input energy within the range of 0 - 75 J.

3.2.2. The Laser Head

The laser head is the laser oscillator which comprised of a laser chamber and laser mirrors. Several laser mirrors with different percentage reflectivities including 75, 90, 95, 97% were employed. The optical components were shown in Figure 3.2. The laser chamber made from ceramic and the comprised of Nd:YAG laser rod and a linear xenon flashlamp.

Figure 3.1. Homemade power supply.

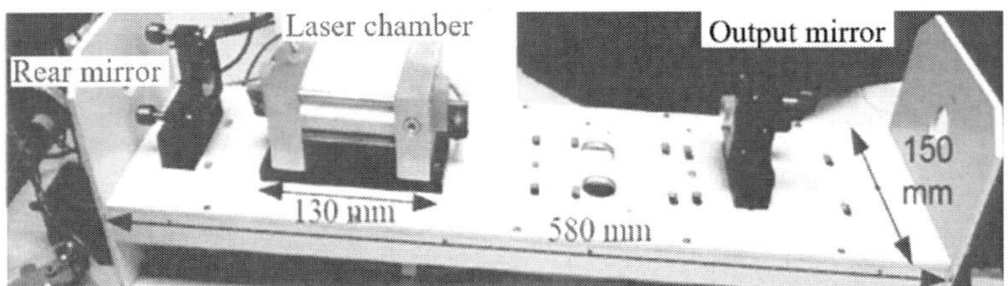

Figure 3.2. Nd:YAG laser head.

3.2.3. The Laser Rod

A Nd:YAG laser rod manufactured from Calstec is employed as an active medium. The doping level of Nd^{3+} in YAG crystal is 1 at.%. The dimension of the rod is 4 mm in diameter and 70 mm in length. Both ends are cut into Brewster angle to minimize the reflection and enhance the polarization beam. The laser rod is enclosed in a samarium flow tube to protect UV light radiation and to keep a steady state of water coolant flow rate. Figure 3.3 indicates, the photo of the Nd:YAG laser rod which held by stainless steel panels.

3.2.4. Optical Resonator

For most lasers, cavity mirrors are vital to increasing the circulating power within the cavity where gain overcomes losses and to increase the rate of stimulated emission for optical amplification and oscillation. Therefore, the optical resonator plays a role in shaping the frequency spectrum of the emitted laser light. This resonator was formed by one flat mirror facing a concave spherical mirror with a radius of curvature larger than the length of the resonator cavity namely Plano-concave resonator. The beam has a relatively large diameter at the spherical mirror and is focused to diffraction limited point at the plane mirror.

The laser resonator was configured as a near-hemispherical resonator having output coupler at the percentage of 75% as shown in Figure 3.4.

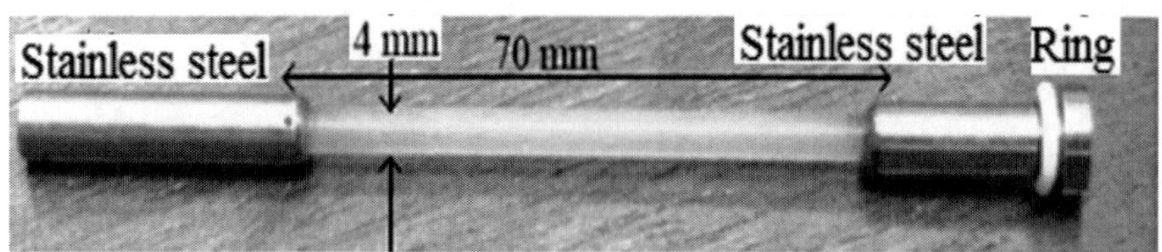

Figure 3.3. Nd:YAG laser rod.

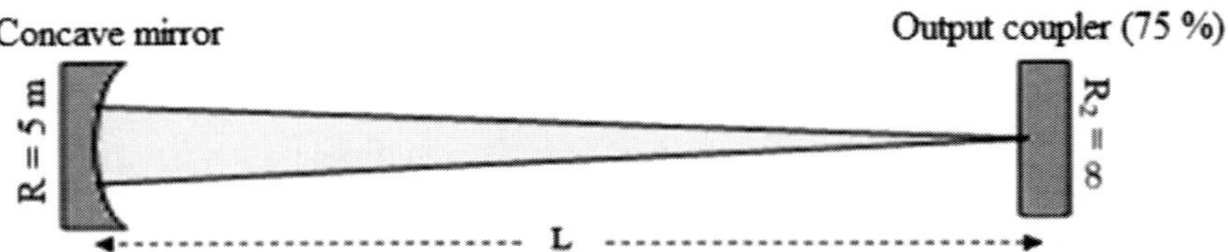

Figure 3.4. Plano-concave resonator configuration.

The optical resonator confined of two mirrors which placed around the gain medium. The mirrors are designed with the precise coating at 946.0 nm to provide feedback of the light in the resonator.

In the present research, the output coupler was flat mirror whereas the rear mirror was a concave mirror with 5 m radius of curvature and both mirrors have 5 mm in thickness and 25.4 mm in diameter.

The laser mirror is indicated in Figure 3.5. The rear mirror with a reflectivity of nearly 100% and output coupler with a reflectance of 75% for 946 nm transition line were used in present work. In section 4.7 characteristics of output coupler and high reflection mirror will be discussed in detail.

Both mirrors were manufactured using BK7 substrates, high reflection coated using dielectric material at 946 nm laser wavelength with high reflectivity of this line, and polished to $\lambda/10$ flatness [1].

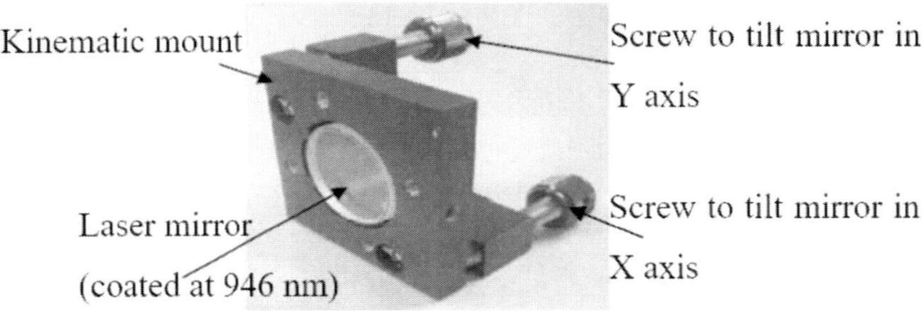

Figure 3.5. Laser mirror on a kinematic mount.

3.2.5. Xenon Flashlamp

A Xenon flashlamp is shown in Figure 3.6. A linear flashlamp filled xenon gas at a pressure of 450 Torr is placed parallel to the laser rod in a ceramic reflector for preparation of side pumping technique. The length and diameter size of the Xenon flashlamp was 75 mm and 4 mm respectively. The impedance parameter, K_0 of the flashlamp was 22.2 Ohm Amp$^{1/2}$. It was enveloped with a transparent fused quartz.

Initially, the flashlamp was ignited with a high voltage of 6 kV to preionize the gas and standby in simmering mode. The capacitor bank with the capacitance of 150 μF was used store and discharge into the flashlamp with input energy within 0 to 75 J. The free running flashlamp was operated in a single and repetitive mode.

The active medium together with the flashlamp was stabilized by water cooling system. For normal the coolant temperature was set at $20°$C and using a mini turbopump for water circulation.

3.2.6. Cooling System

Xenon flashlamp radiation generally contains a wide range of electromagnetic spectrum from ultraviolet to near IR wavelength. Since only some wavelengths of light are absorbed, other wavelengths of light emerge as heat. As a result thermal stress is produced on the laser rod. Laser performance is affected by this thermal stress and laser efficiency of the laser performance will decrease and flashlamp, laser rod, and pump cavity may be damaged. Chatude

Thus the Nd:YAG crystal together with the flashlamp was flooded with a coolant comprised of the mixture of 60% ethylene glycol and 40% distilled water. Such particular coolant covers a range of temperature from -30°C to +60°C. A turbo pump was provided to fast circulate the water into the laser head. Figure 3.7 shows the picture of the cooling system.

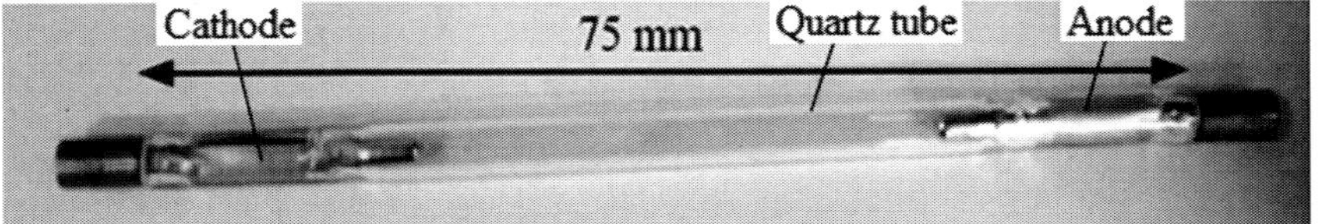

Figure 3.6. A linear xenon flashlamp.

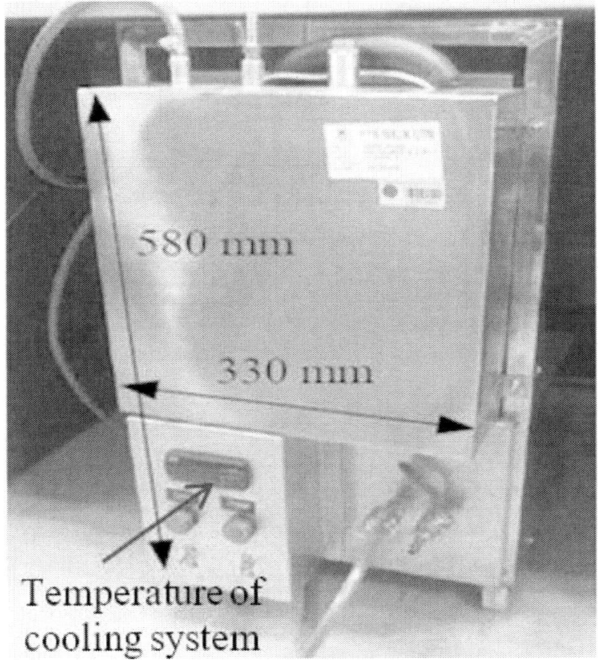

Figure 3.7. Cooling system and temperature controller.

3.2.7. Ethylene Glycol

The major use of ethylene glycol is a medium for convective heat transfer. Ethylene glycol based water solution is common in heat-transfer applications where the temperature in the heat transfer fluid can be below 0°C. Ethylene glycol is also used in systems that must cool below the freezing temperature of water. A mixture of 60% ethylene glycol and 40% water freezes at -48°C.

Table 3.1. Ethylene glycol freezing and boiling point versus concentration in water

Weight percentage	0	10	20	30	40	50	60	70	80	90	100
Freezing point	0	-4	-7	-15	-23	-34	-48	-51	-45	-29	-12
Boiling Point	100	102	102	104	104	107	110	116	124	140	197

However, the boiling point for ethylene glycol increases monotonically with increasing ethylene glycol percentage. Thus the use of ethylene glycol not only depresses the freezing point but also elevates the boiling point. Table 3.1 summarizes the detailed percentage of ethylene glycol for freezing and boiling point.

3.2.8. Spectrometer

The fluorescence radiation was emitted at one end of the laser rod and the wavelengths in the range of 550 nm to 1100 nm detected by an Ophir Wavestar which is shown in Figure 3.8. The detection device comprised a photodetector which is sensitive in the region of near Infrared. The emission spectrum is analyzed via Wavestar version 1.05 software. The resolution of this detection system is 0.5 nm so it can resolve most of the transition lines appeared from the pumping rod. The spectrometer was interfaced to a personal computer.

Figure 3.8. Ophir spectrometer

3.2.9. Output Energy Measurement

Single pulse and variable repetition rates of output energy of the Nd^{3+}:YAG laser were investigated by evaluating their pulse energy. Figure 3.9 indicates the experimental setup to measure the long pulse output laser.

Mellios-Griot energy/power meter model 13PEM 001/J was employed to measure the output energy of the laser as shown in Figure 3.9. A thermopile sensor responds to a wide range of electromagnetic radiation from ultraviolet to infrared. By increasing the operation voltage of CCPS power supply in the range of 300 V to 1000 V variable laser output energy can be achieved.

3.3. EXPERIMENTAL METHOD

The schematic diagram of the experimental set-up is shown in Figure 3.10. The system comprises of two parts. The first part is energizing the laser crystal and stabilizing the stimulating emission of fluorescence radiation. The second part is a measurement of spectroscopy properties of both flashlamp and Nd:YAG laser rod.

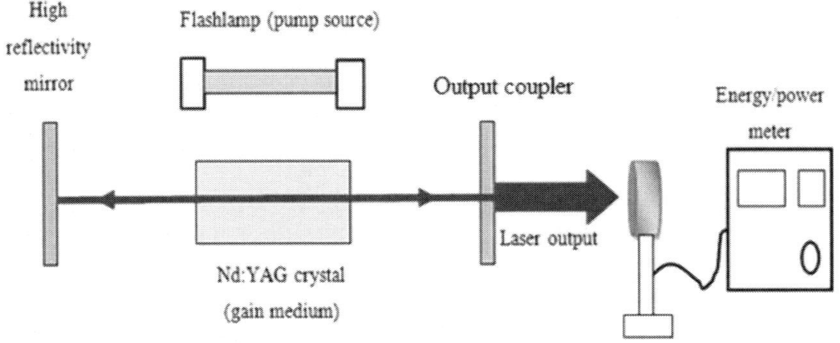

Figure 3.9. Experimental setup of output energy measurement.

The laser rod is placed parallel to a linear flashlamp filled xenon gas. The flashlamp is pumped by a homemade power supply. The driver is triggered by simmer mode technique. A capacitor bank with the capacitance of 150 μF is charged by the maximum voltage of 1000 V thus the input energy is varied between 0 to 75 J.

The Nd:YAG crystal together with the flashlamp is flooded with a coolant comprised of the mixture of 60% ethylene glycol and 40% distilled water. Such particular coolant covers the range of temperature from -30°C to +60°C. The rod is further enclosed with a samarium flow tube to absorb UV light radiation and to keep the flow rate in a steady state condition. The temperature of coolant can be varied by manipulating the buttons that devised in the cooling system.

Figure 3.10 shows the experimental setup before alignment the mirrors. The output spectrum lines were detected via a CCD camera placed at 20 cm from the source and interfaced with a computer. The emission spectrum is analyzed via a Wavestar version 1.05 software and the intensity, wavelength, and linewidth of each line of the spectrum radiation were recorded.

Figure 3.11 shows the experimental setup to characterize the optical resonator mirrors. By separate installation of both high reflectivity and output, coupler mirrors the spectrum lines of flashlamp and Nd:YAG rod while they passed through the mirrors was recorded. By comparing the intensities of Nd:YAG emission lines before and after alignment of the mirrors characteristics of the rare mirror (high reflectivity) and output coupler (partial reflectivity) can be obtained.

Figure 3.12 shows the experimental setup after alignment the laser mirrors. There were two laser mirrors each coating at 946.0 nm to provide feedback on this line. One mirror acted as high reflection mirror and the other one was an output coupler with 75% reflectivity for 946.0 nm transition line. Both mirrors were properly aligned to form a linear cavity with a length of 30 cm. Laser lines after passing the output coupler mirror were recorded by the spectrometer and analyzed by wave star software.

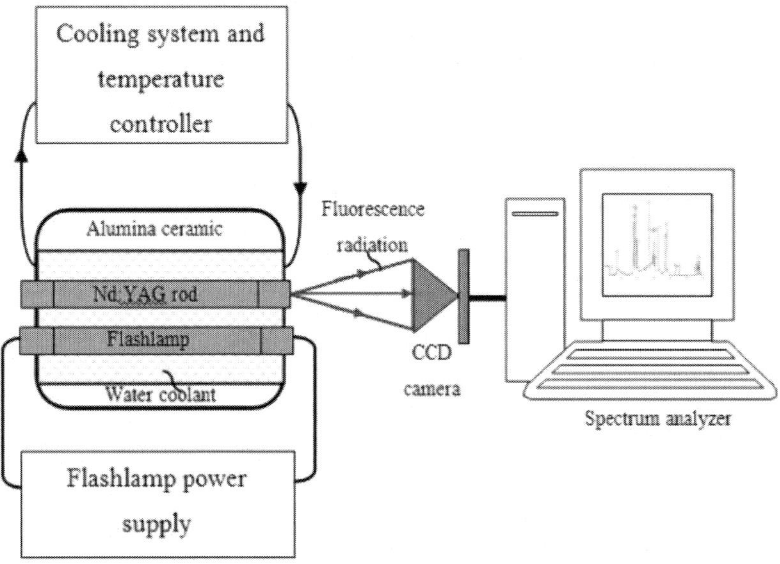

Figure 3.10. Experimental setup before alignment the mirrors.

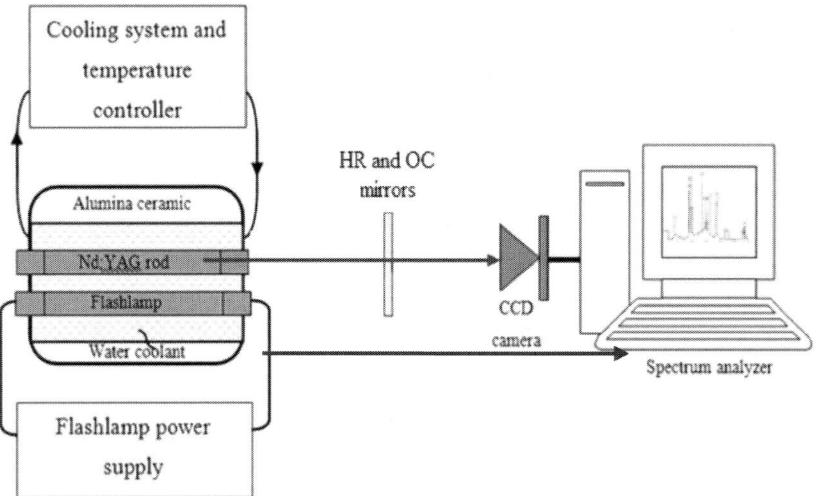

Figure 3.11. Experimental setup to characterize the optical resonator mirrors.

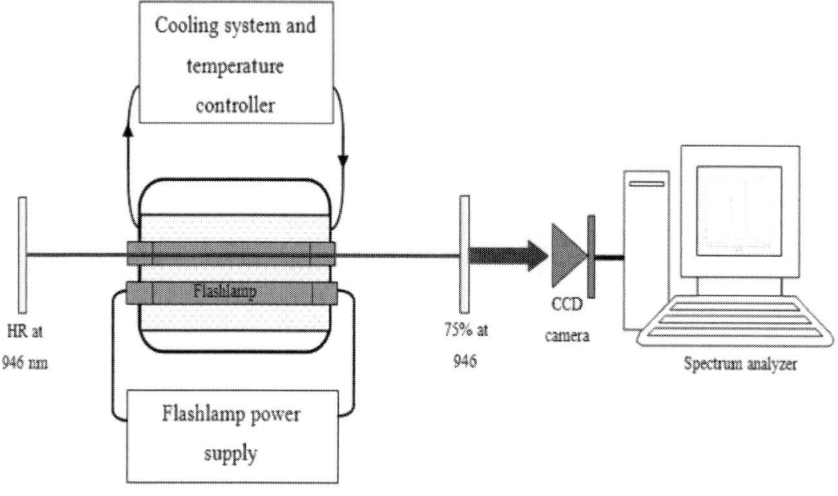

Figure 3.12. Experimental setup after alignment the mirrors to provide feedback of 946 nm transition line.

REFERENCE

[1]   Zainal, R. (2011). *Design and construction of free running and variable repetitive rate of flashlamp pumped Nd:YAG Laser.* PhD thesis, Universiti Teknologi Malaysia.

Chapter 4

Spectroscopic Results and Discussion on the Fluorescence Spectrum of the Pump Source to Excite Nd:YAG Laser Rod as Flashlamp

Abstract

In this chapter, the results of the experiment will be demonstrated, analyzed and discussed. The results are including the spectroscopic study on the fluorescence spectrum of the pump source used in this work to excite Nd:YAG laser rod that is flashlamp.

Keywords: spectroscopic results, Nd3+:YAG laser crystal, flashlamp

4.1. Experimental Demonstrations of Nd:YAG Laser

The flashlamp is powered by an electrical driver at various input voltage from 0 - 1000 V. This followed by the investigation on the spectroscopic study on the fluorescence of the Nd:YAG laser rod after pumped by flashlamp radiation source. The fluorescence was observed in the two intermanifold Stark energy level transitions that are in quasi-three-level laser system and four-level system. Finally, the laser chamber is confined between two mirrors coating at 946.0 nm to study the spectroscopic of laser radiation at quasi-three-level laser performance. In this chapter, the results of the experiment will be demonstrated, analyzed and discussed. The results are including the spectroscopic study on the fluorescence spectrum of the pump source used in this work to excite Nd:YAG laser rod that is flashlamp.

4.2. SPECTROSCOPIC OF XENON FLASHLAMP

In this section, the fluorescence radiation of flashlamp filled by xenon gas is demonstrated. The selection of flashlamp as a pumping source is important to ensure the production of radiation spectrum matched with the absorption band of Nd:YAG laser crystal.

The typical result obtained by this experiment is shown in a series of figure including Figure 4.1 to Figure 4.5 which the fluorescence radiation covered in the range of 540 – 1080 nm. The spectra comprised of atomic lines (the right side of the figure) and continuum components (the left side of the figure). The atomic lines correspond to the discrete transition between the bound energy states of the gas atoms and ions which were also known as bound-bound transitions. While the continuum part is primarily due to recombination radiation from gas ions capturing electrons into bound state and bremsstrahlung radiation from electrons accelerated during collisions with ions.

The intensity of longer wavelength at low density is prominent than the shorter wavelength. As the current density increases by charging the capacitor from 300 to 700 V, as shown in Figures 4.1 to 4.5, the continuum part is almost dominant that implying shorter wavelengths have greater intensity than longer wavelengths.

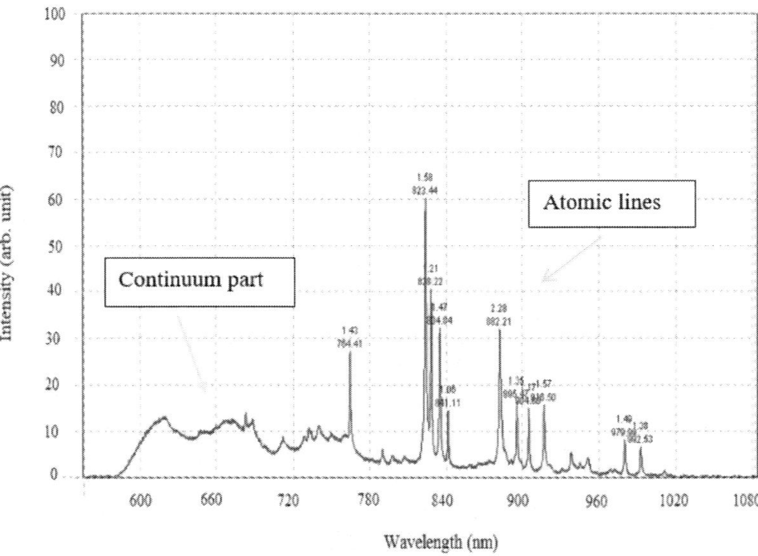

Figure 4.1. Spectral emissions from Xenon flashlamp (150 µF capacitors charged) at a voltage of 300 V.

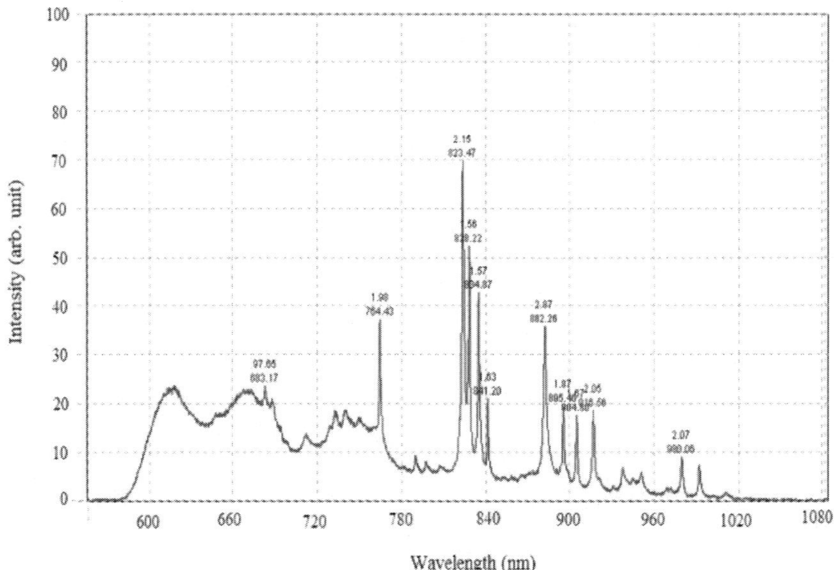

Figure 4.2. Spectral emissions from Xenon flashlamp (150 µF capacitors charged) at a voltage of 400 V.

        Seyed Ebrahim Pourmand, Iraj Sadegh Amiri et al.

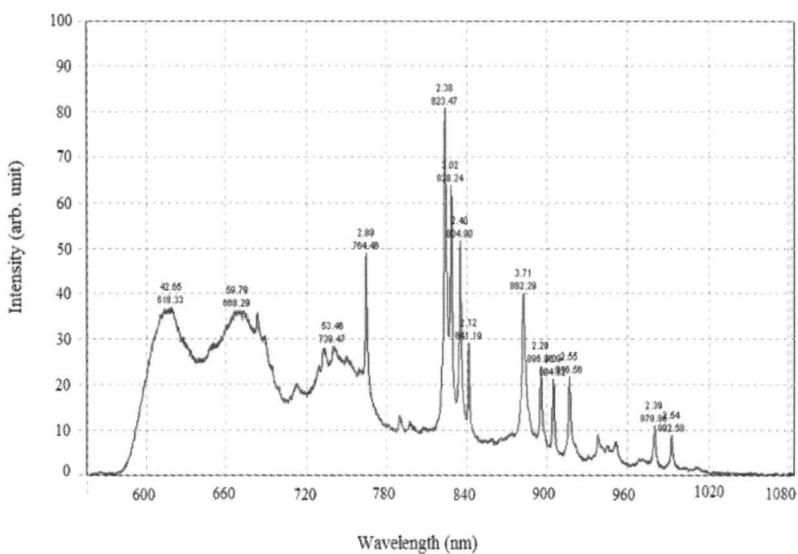

Figure 4.3. Spectral emissions from Xenon flashlamp (150 µF capacitors charged) at a voltage of 500 V.

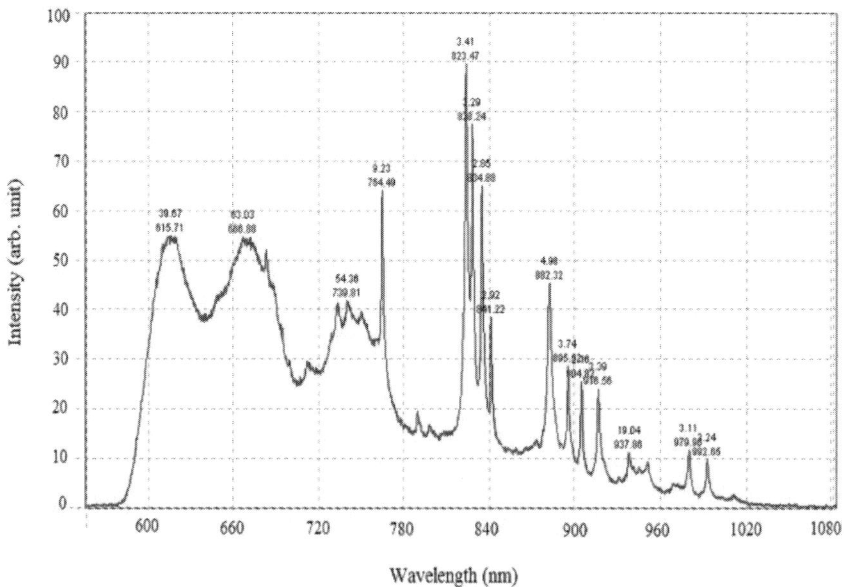

Figure 4.4. Spectral emissions from Xenon flashlamp (150 µF capacitors charged) at a voltage of 600 V.

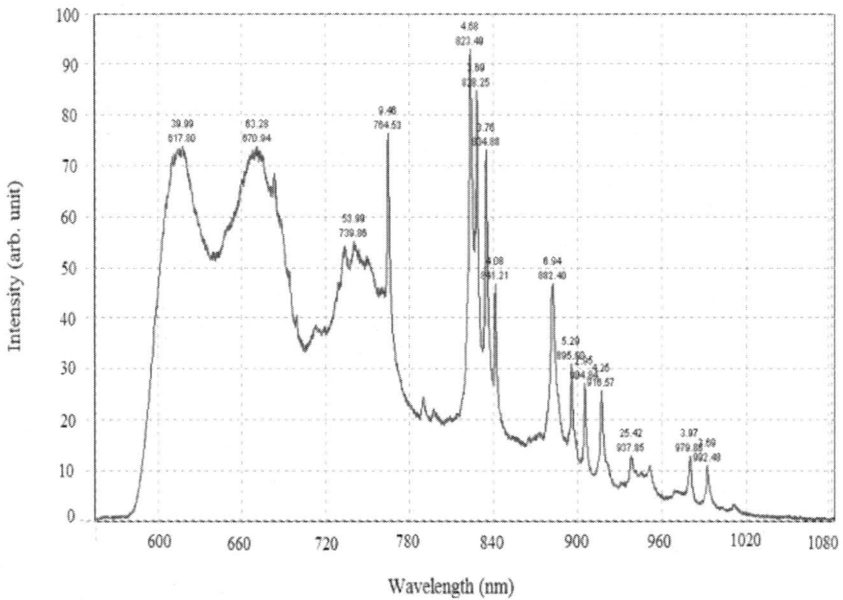

Figure 4.5. Spectral emissions from Xenon flashlamp (150 µF capacitors charged) at a voltage of 700 V.

Figure 4.6 represents the growing trend of the emission intensity of continuum part at 615 nm and atomic lines of 764.49, 823 .47, 828.24, 834.88, 841.22 and 882.32 nm at various voltage.

The main absorption band of Nd:YAG is at 0.525 to 0.585 µm, 0.73 to 0.75 µm and 0.79 to 0.9 µm as illustrated in Figure 3.7. The main absorption lines are including 764.41, 823.44, 828.22, 834.84, 841.11, 882.21 and 895.47 nm which are responsible to excite Nd^{3+} from ground state $^4I_{9/2}$ to pump states above the laser level state at $^4F_{7/2}$, $^4F_{5/2}$, $^4F_{5/2}$, $^4H_{9/2}$, $^4F_{3/2}$ respectively.

Experimental results depict that the transition line at 615.71 nm (continuum part) has the most gradual slope. In high current densities due to increasing the voltage the spectral output of the flashlamp is dominated by continuum radiation, and the line structure is seen as a relatively minor element. This phenomenon is arising from the behavior of gas which acts as a blackbody radiation at high densities.

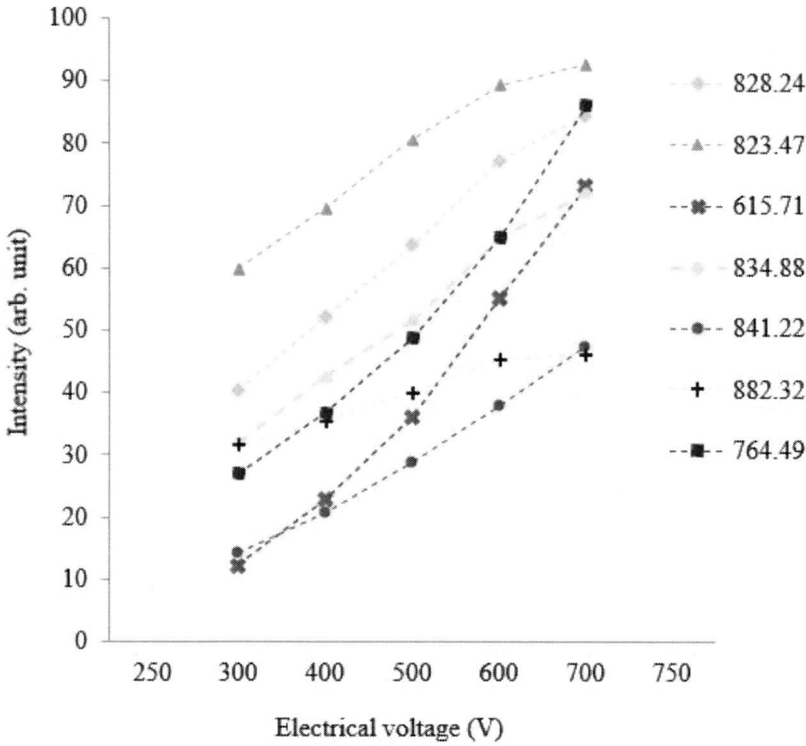

Figure 4.6. Intensity of several lines of flashlamp radiation.

4.3. EMISSION CROSS SECTION OF INTER STARK TRANSITIONS

The spectra of Nd:YAG laser rod pumped by flashlamp is shown in Figure 4.7 which includes quasi-three-level and four level transitions. It realized that quasi-three-level emissions are far small than four level emissions when operating at room temperature.

The most important parameters from these spectral comprise of intensity and linewidth. The knowledge of intensity for spectral lines allows the computational of branching ratio. Notice that branching ratio is defined as the intensity of a certain atomic line divided with the total intensity of the fluorescence lines that emitted after pumping with flashlamp, which included the quasi-three-level laser and four laser system transition lines. Some of the spectroscopic properties that are important to determine the performance of Nd:YAG laser pumped by flashlamp are listed in Table 4.1.

The measurement of linewidth from each of the transition lines in the spectrum are used for computing the emission cross section σ_E with Eq. (2.13). The calculated results at 20°C were summarized in Table 4.1. The average emission cross section at 946 and 1064 nm are 0.69 × 10^{-19} cm^2 and 2.94 × 10^{-19} cm^2 respectively. The values obtained agree well compared to the other researcher.

In the following sections, the effects of temperature and input energy on some critical parameters such as linewidth, line shift, intensity and emission cross section will be investigated.

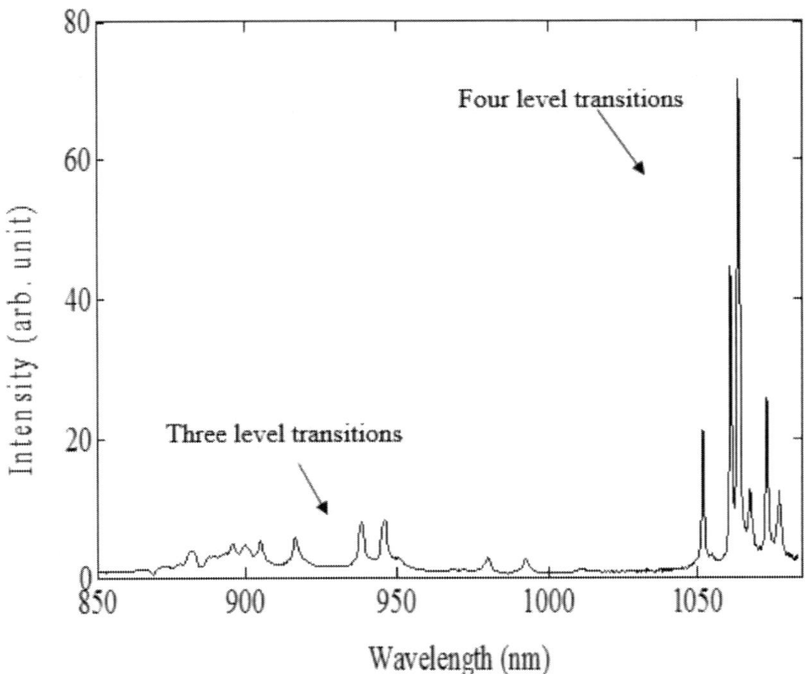

Figure 4.7. Fluorescence spectrum of flashlamp pumped Nd:YAG at 300K.

Table 4.1. Spectroscopy properties of some inter Stark transition lines

Wavelength, λ (nm)	Branching Ratio, BR	Average Linewidth, $\Delta\lambda$ (nm)	Cross section, $\sigma_E \times 10^{-19}$ (cm^2)	Saturation Intensity, $I_s \times 10^3$ (Wcm^{-2})
938.44	0.252	1.98	0.47	18.0
946.07	0.316	1.74	0.69	12.2
1051.86	0.095	1.09	0.96	7.9
1061.46	0.199	0.99	2.43	3.1
1063.94	0.319	1.33	2.94	2.5
1073.76	0.115	1.14	1.28	5.7
1077.47	0.056	2.25	0.73	10.1

4.4. SPECTROSCOPY PROPERTIES OF QUASI THREE-LEVEL AND FOUR LEVEL TRANSITION LINES AT DIFFERENT INPUT ENERGIES

The fluorescence spectra of the main lines of the manifold transitions $^4F_{3/2}$ to $^4I_{11/2}$ at 1% Nd^{3+} in YAG versus input energy in the range of 18 to 75 J (corresponding to the voltage from 500 to 100 V) are shown in Figure 4.8 to Figure 4.13.

In general, Figures 4.8 to 4.13 show the quasi-three-level transitions comprised of eight lines, 867.02, 874.22, 878.02, 885.01, 890.25, 900.09, 938.51 and 946.08 nm. The spectroscopy properties of each line change with input energy. Three important spectroscopy parameters including the intensity, linewidth, and line shift were investigated. The obtainable data from the figures were collected and tabulated in Table 4.2 – 4.4. Among line spectra, only the last two lines obviously still manage to maintain the linewidth below 3 nm. Therefore the major line in quasi-three-level lasers was taken as 938 and 946 nm.

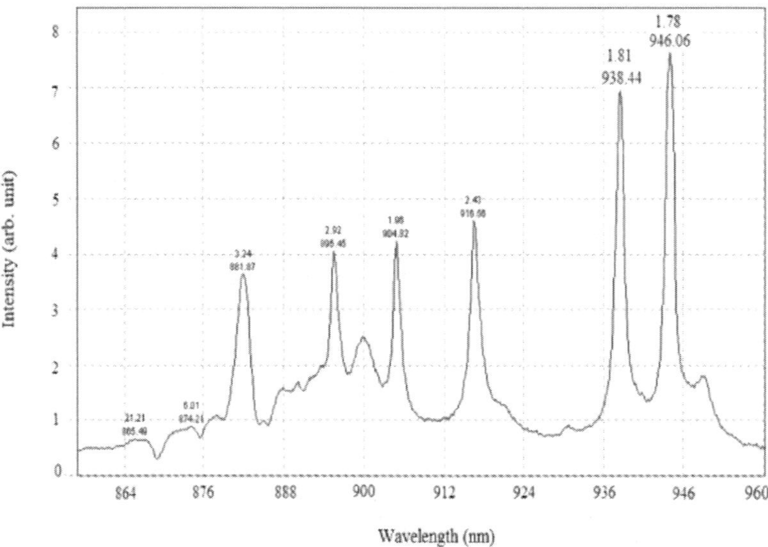

Figure 4.8. Spectroscopic properties of quasi-three-level system emissions pumped at input energy of 18.75 J.

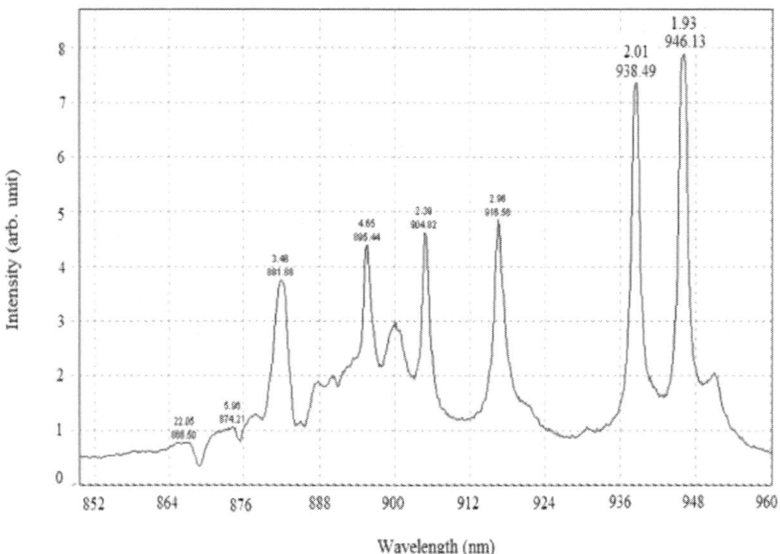

Figure 4.9. Spectroscopic properties of quasi-three-level system emissions pumped at input energy of 27.00 J.

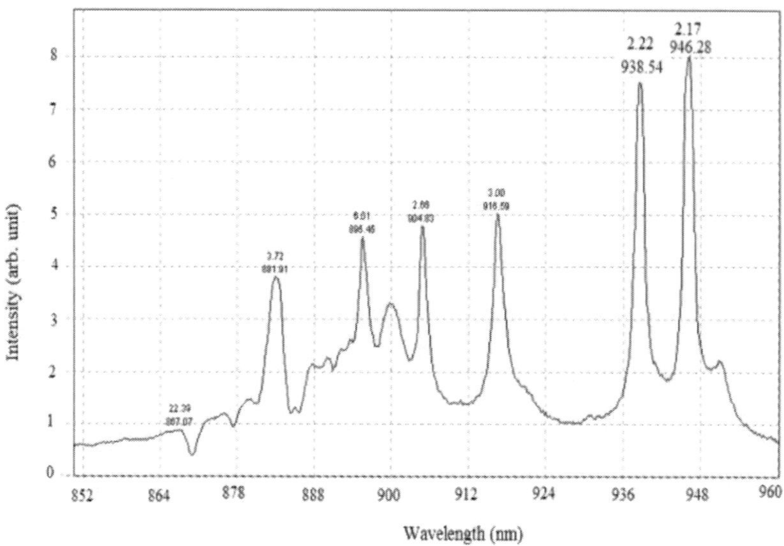

Figure 4.10. Spectroscopic properties of quasi-three-level system emissions pumped at input energy of 36.75 J.

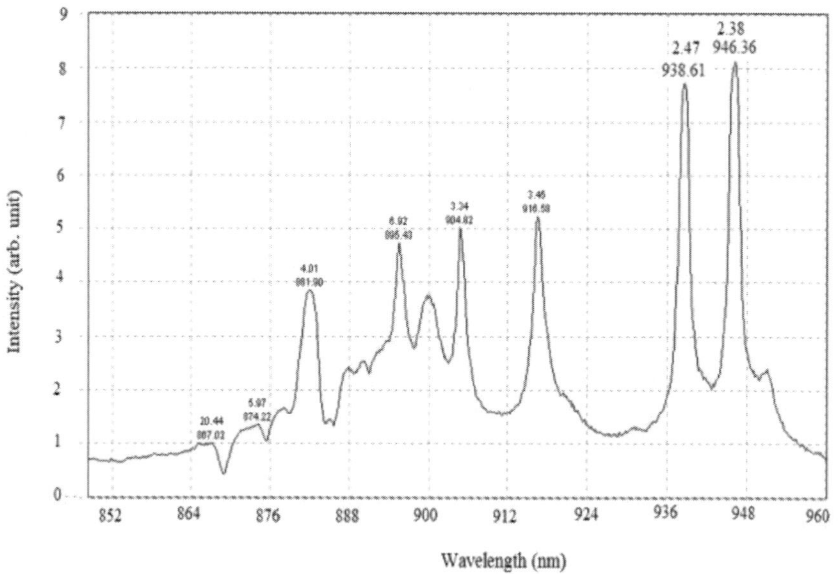

Figure 4.11. Spectroscopic properties of quasi-three-level system emissions pumped at input energy of 48.00 J.

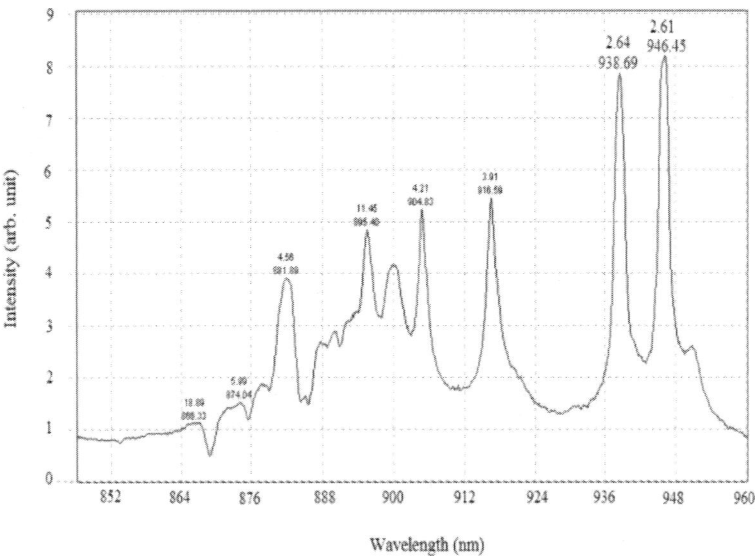

Figure 4.12. Spectroscopic properties of quasi-three-level system emissions pumped at input energy of 60.75 J.

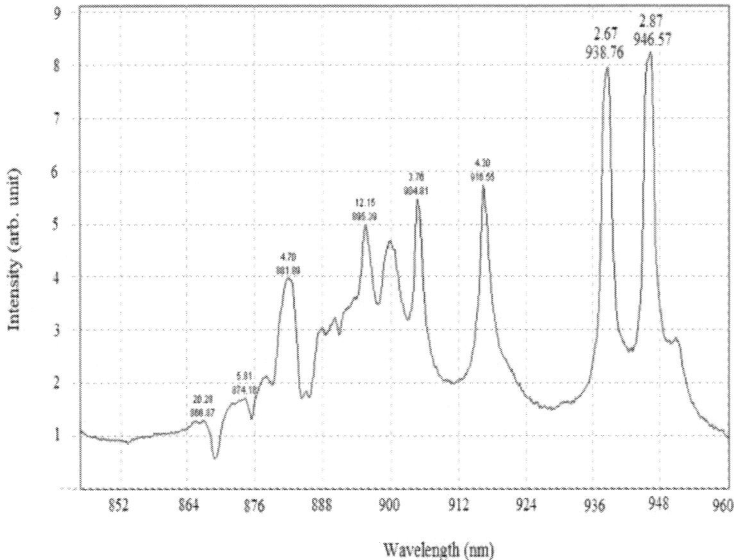

Figure 4.13. Spectroscopic properties of quasi-three-level system emissions pumped at input energy of 75.00 J.

**Table 4.2. Spectroscopy properties of quasi-three-level
lines – Linewidth variation**

Spectra (nm)	Linewidth at different input energy, (nm)					
	18.75 J	27.00 J	36.75 J	48.00 J	60.75 J	75.00 J
938.44	1.81	2.01	2.22	2.47	2.64	2.67
946.06	1.78	1.93	2.17	2.38	2.61	2.87

**Table 4.3. Spectroscopy properties of quasi-three-level lines –
Intensity variation**

Spectra (nm)	Intensity at different energy (a.u)					
	18.75 J	27.00 J	36.75	48.00 J	60.75	75.00 J
867.02	0.61	0.73	0.86	0.96	1.10	1.26
874.22	0.87	1.01	1.16	1.33	1.50	1.69
878.02	1.07	1.25	1.45	1.66	1.85	2.11
885.01	0.96	1.12	1.29	1.43	1.64	1.82
890.25	1.66	1.96	2.25	2.53	2.88	3.21
900.09	2.49	2.96	3.27	3.72	4.19	4.66
938.51	6.93	7.32	7.52	7.70	7.83	7.94
946.08	7.61	7.87	8.02	8.13	8.17	8.22

**Table 4.4. Spectroscopy properties of quasi-three-level lines – Line
shift variation**

Line shift (nm)	Input energy					
	18.75 J	27.00 J	36.75	48.00 J	60.75 J	75.00 J
938	938.44	938.49	938.54	938.61	938.69	938.76
946	946.06	946.13	946.28	946.36	946.45	946.57

The typical results obtained at four level transitions from manifold
$^4F_{3/2}$ to $^4I_{11/2}$ are shown in Figure 4.14 – 4.19. There are six lines can be
detected by the detector used in this experiment. There are including
1051, 1061, 1064, 1068, 1073 and 1075 nm. All the lines experience a
change in the spectroscopy properties including intensity, linewidth,
and line shift. Since this work just concerns only for the three-level

laser system, therefore these four-level lines are important in order to get the total intensity involve in the fluorescence radiated emitted after the laser rod was pumped by the flashlamp.

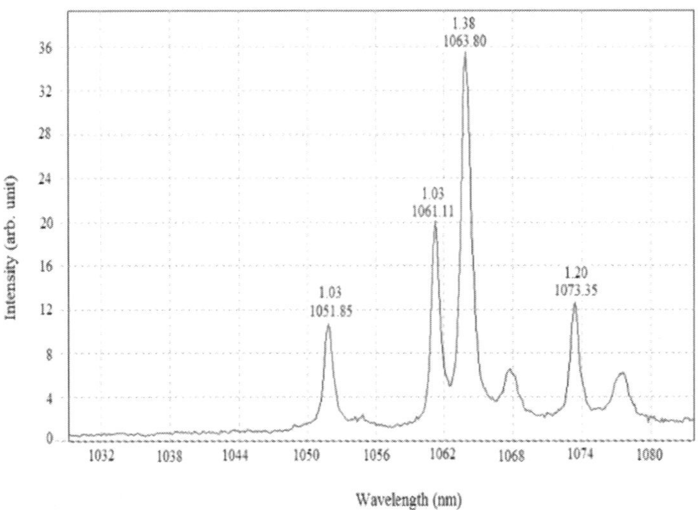

Figure 4.14. Spectrum of four level laser transitions at input energy of 18.75 J.

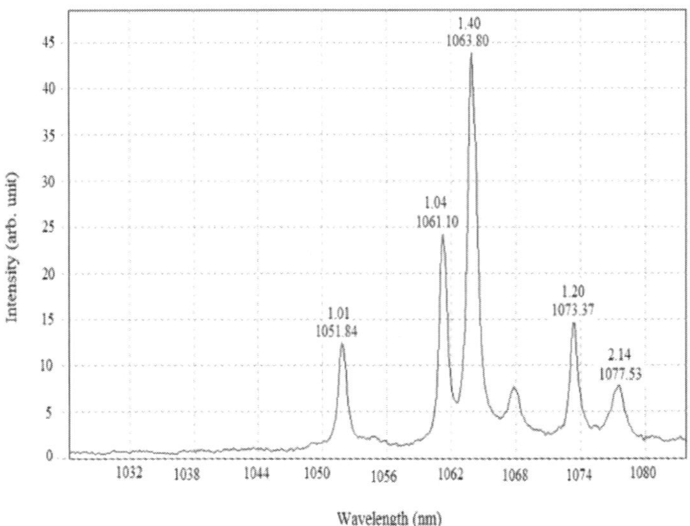

Figure 4.15. Spectrum of four level laser transitions at input energy of 27.00 J.

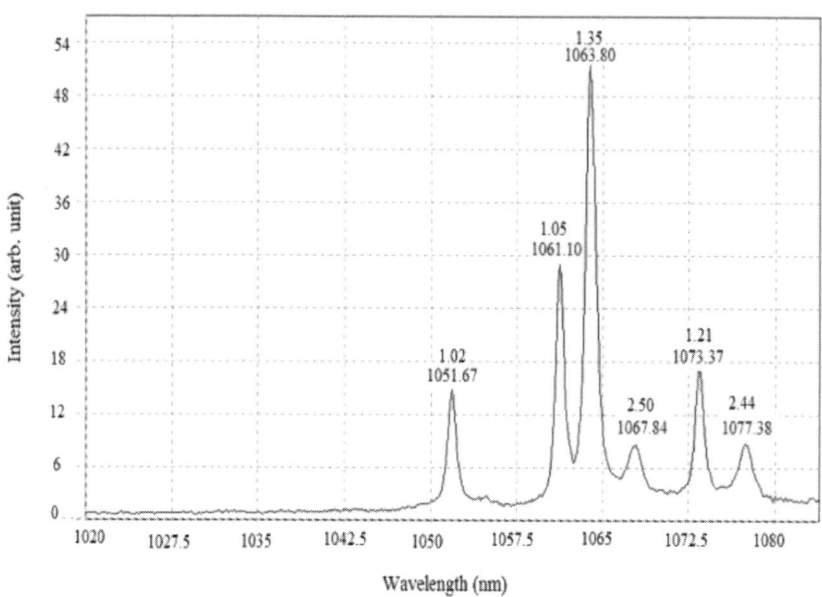

Figure 4.16. Spectrum of four level laser transitions at input energy of 36.75 J.

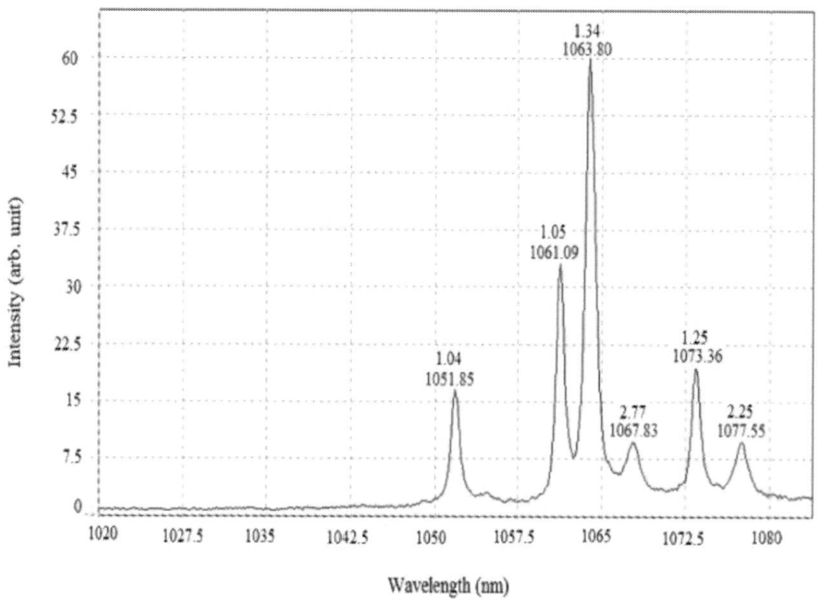

Figure 4.17. Spectrum of four level laser transitions at input energy of 48.00 J.

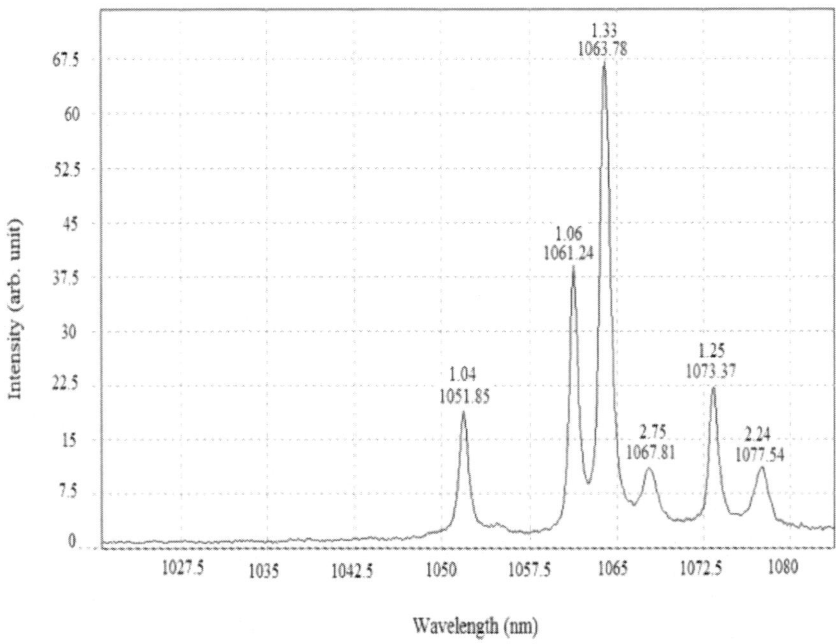

Figure 4.18. Spectrum of four level laser transitions at input energy of 60.75 J.

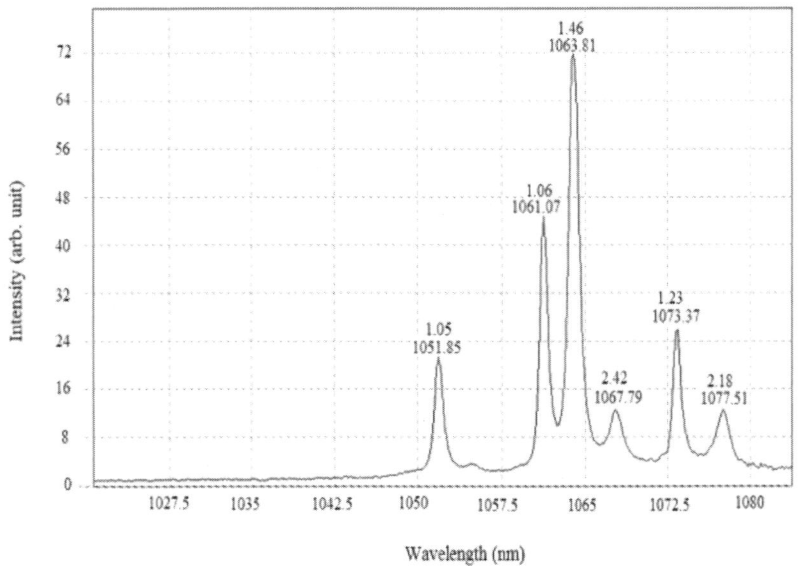

Figure 4.19. Spectrum of four level laser transitions at input energy of 75.00 J.

 Seyed Ebrahim Pourmand, Iraj Sadegh Amiri et al.

**Table 4.5. Spectroscopy properties of four level transitions –
Linewidth variation**

Spectra (nm)	Linewidth at different input energy					
	18.75 J	27.00	36.75	48.00	60.75	75.00
1051.85	1.03	1.01	1.02	1.04	1.04	1.05
1061.11	1.03	1.04	1.05	1.05	1.06	1.06
1063.81	1.38	1.40	1.35	1.34	1.33	1.46
1067.79	-	-	2.50	2.77	2.75	2.42
1073.37	1.20	1.20	1.21	1.25	1.25	1.23
1077.51	-	2.14	2.44	2.25	2.24	2.18

**Table 4.6. Spectroscopy properties of four level transitions –
Intensity variation**

Spectra (nm)	Intensity at different energy (a.u)					
	18.75 J	27.00 J	36.75 J	48.00 J	60.75 J	75.00 J
1051.85	10.60	12.34	14.64	16.61	18.73	21.23
1061.11	20.04	24.18	28.93	32.95	38.87	44.55
1063.80	35.40	43.74	51.40	59.65	66.98	71.55
1067.84	6.42	7.62	8.36	9.63	10.90	12.55
1073.35	12.55	14.61	16.92	19.31	21.92	25.83
1077.53	6.19	7.79	8.52	9.69	10.91	12.48

**Table 4.7. Spectroscopy properties of four level transitions – Line
shift variation**

Spectra (nm)	Line shift at different input energy					
	18.00 J	27.00 J	36.75 J	48.00 J	60.75 J	75.00 J
1051.85	1051.85	1051.84	1051.67	1051.85	1051.85	1051.85
1061.11	1061.11	1061.10	1061.10	1061.09	1061.24	1061.07
1063.80	1063.80	1063.80	1063.80	1063.80	1063.78	1063.81
1067.84	-	-	1067.84	1067.83	1067.81	1067.79
1073.35	1073.35	1073.37	1073.37	1073.36	1073.37	1073.37
1077.53	-	1077.53	1077.38	1077.55	1077.54	1077.51

The linewidth, intensity and line shift of four level transition lines from Figures 4.14 to 4.19 were collected and tabulated in Table 4.5, 4.6 and 4.7 respectively.

Xenon flashlamp radiation generally covers a wide range of electromagnetic spectrum from ultraviolet to near IR wavelength. Since only some wavelengths of light are absorbed, other wavelengths of light emerge as heat on the laser rod. This thermal effect can dramatically influence the laser performance.

However, with accurate information on thermal effects, it would be possible to deal with the varying stability of the output energy over the pumping energy range of interest. In addition, thermal broadening and line shifts of transition lines in crystals yield precious knowledge on the spectrum of vibrational excitations and also on the electron-phonon interactions.

The input energy dependence of positions of 938.5 and 946.0 nm lines and 1051.9, 1061.3, 1063.9, 1067.6, 1073.3 and 1077.5 nm emission lines at input energy from 18 J to75 J are shown in Figure 4.20. The experimental results show that the positions of four level system emission lines of Nd^{3+}:YAG crystal remain constant with input energy.

Entirely different results are obtained with the quasi-three-level emission lines whereby obviously they are apparently an input energy dependent. The 938.5 and 946 nm lines in intermanifold $^4F_{3/2}$ to $^4I_{9/2}$ shifted to longer wavelength (red shift) and as a result, the energy level of Stark sublevels of Z_5 and Z_4 move upward. Figure 4.20 shows that the main emission lines of quasi-three-level are observed to shift from 938.44 nm and 946.06 nm at 18 J to 938.76 nm and 946.57 nm at 75.00 J respectively.

Figure 4.21 indicates the linewidths dependence of 938.5 and 946.0 nm emission lines along with 1051.9, 1061.3, 1063.9, 1067.6, 1073.3 and 1077.5 nm emission lines at input energy from 18 J to 75 J.

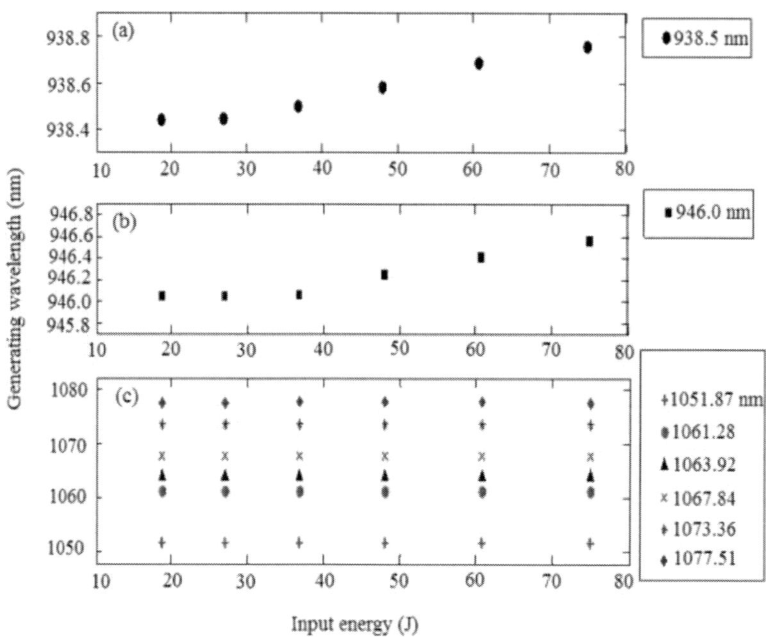

Figure 4.20. Input energy dependence of (a) line of 938.5 nm, (b) line of 946 nm and (c) several four level emission lines.

In addition, all linewidths of transition lines from intermanifold $^4F_{3/2}$ to $^4I_{11/2}$ (four-level system) is found to remain constant with input energy. In the case of a quasi three-level system, as the input energy increases, the spectral linewidths increase. The input energy in the range of 18-75 J, the linewidth increased by 103% and 93% for the $R_1 \rightarrow X_5$ and $R_2 \rightarrow X_5$ transitions respectively.

Since there is no linewidth and position change at four level system transition lines, it can be assumed that there is no any ion phonon interaction such as direct one phonon processes $R_1 \rightarrow R_2$ and $Y_1 \rightarrow Y_2$, Y_3 ... Y_6 and multi-phonon emission processes from Stark level R_1 and R_2 to $^4F_{3/2}$ manifold by increasing the input energy. In addition energy widths of R_1 and R_2 and Y_1, Y_2 ... Y_6 also remain unchanged over the studied range. In addition, the shape of the spectral lines of the 938.5 and 946.0 nm could be fitted by Lorentzian at room temperature, therefore the line broadening arising from the crystal strain inhomogeneity can be neglected.

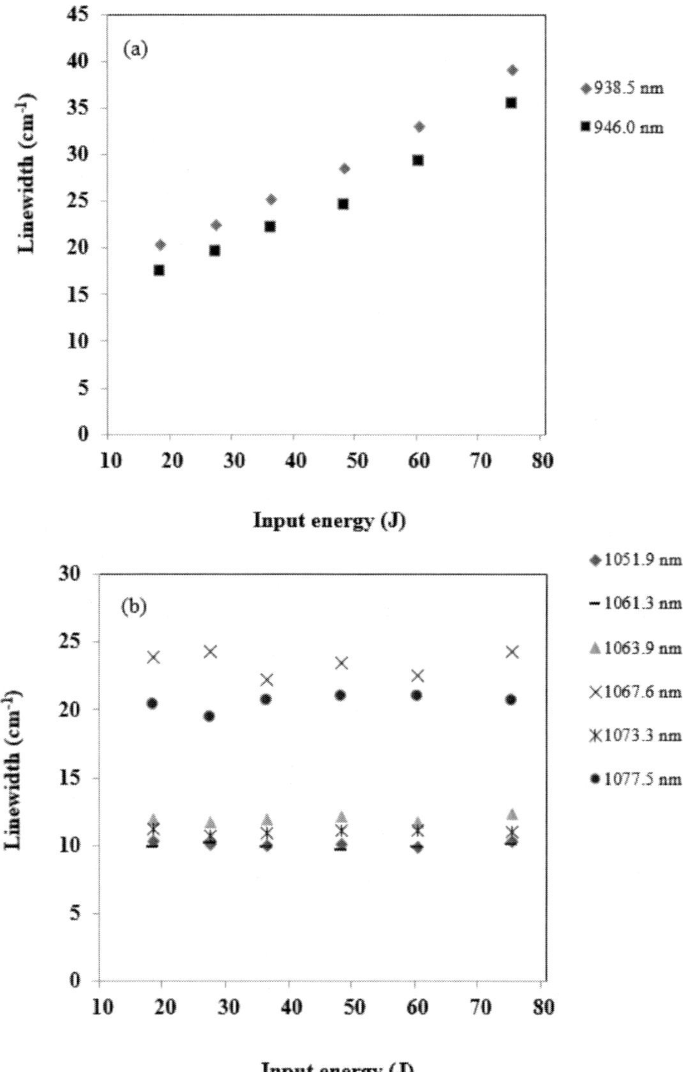

Figure 4.21. Linewidth of (a) 938.5 and 946 nm lines and (b) several four-level system lines versus input energy.

Besides, the present work is done at a constant temperature of 20°C, thus only input energy dependence of linewidth and position of transition lines are considered. Thus, Raman scattering and one-phonon temperature dependent part are neglected. Hence the thermal broadening of the quasi-three-level emission lines are mainly related to the temperature independent of spontaneous one-phonon emission and multi-phonon processes.

Furthermore, increasing the input energy causes heavy thermal population in the ground sublevels. As a result spontaneous one phonon emission $X_5 \rightarrow X_4$, X_3, . . ., X_1 processes will increase and significantly affect online shift and linewidth of inter stark transitions which terminate to the ground level.

Finally, Figure 4.21 shows that the $R_1 \rightarrow Y_1$ transition has the smallest linewidth between four-level transition lines. This is owing to least direct one-phonon and multi-phonon processes contribute in this emission line. In the intermanifold of $^4F_{3/2}$ and $^4I_{11/2}$, the Stark sublevels of R_1 and Y_1 have the lowest energy levels, therefore, spontaneous one-phonon emission processes are unlikely in the transition line of $R_1 \rightarrow Y_1$.

However, to the best of our knowledge, input energy dependency of linewidth and wavelength position of quasi-three-level and four level system transitions which were investigated at the present work has not been reported up to now.

4.5. SPECTROSCOPY PROPERTIES OF QUASI THREE LEVEL TRANSITIONS AT DIFFERENT TEMPERATURE

The typical result of spectrograph of Nd:YAG fluorescence at a different temperature is shown in Figure 4.22 – 5.25. All spectrographs are arranged in the increasing order of the temperature. The important information collected from these figures was summarized in Table 4.8, 4.9 and 4.10.

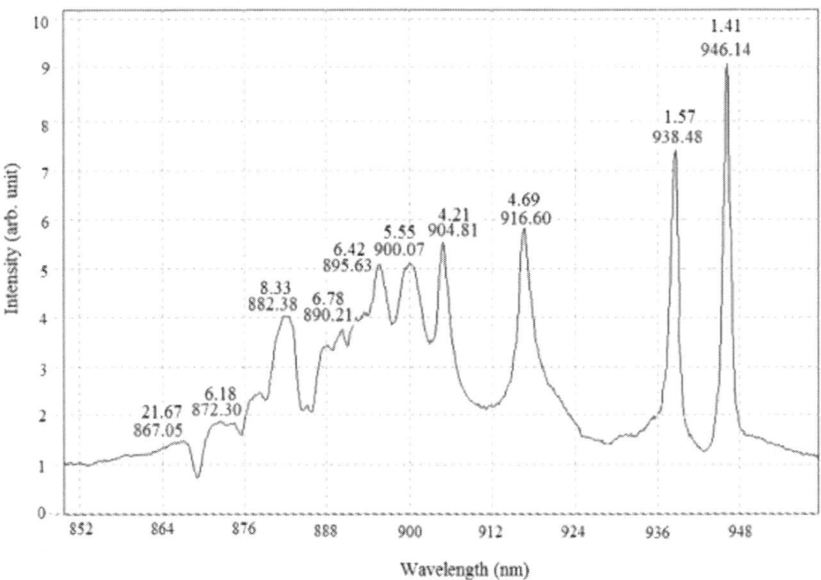

Figure 4.22. Spectrograph of quasi-three-level laser at temperature of -30°C.

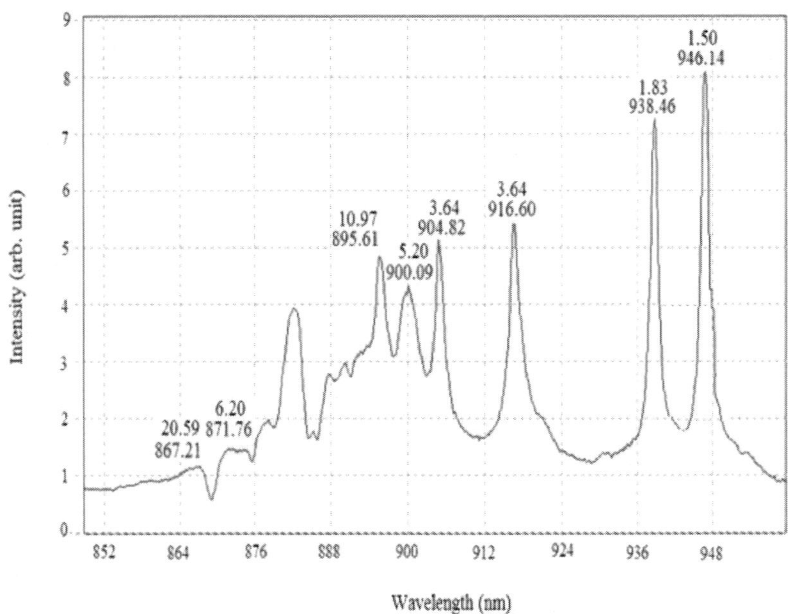

Figure 4.23. Spectrograph of quasi-three-level laser at temperature of 0°C.

Seyed Ebrahim Pourmand, Iraj Sadegh Amiri et al.

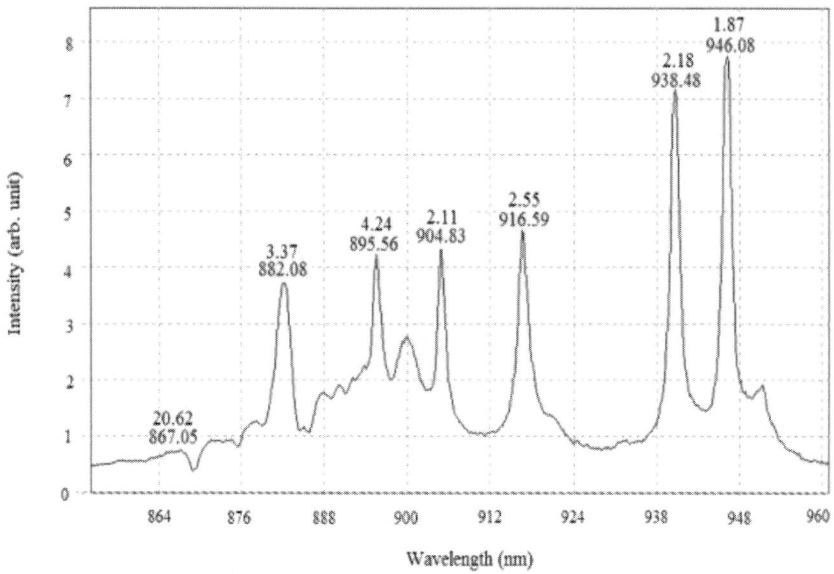

Figure 4.24. Spectrograph of quasi-three-level laser at temperature of 30°C.

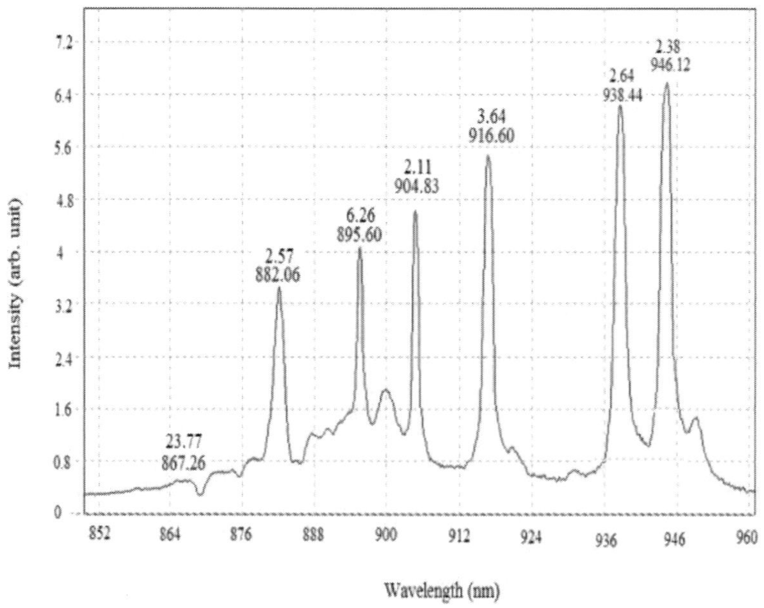

Figure 4.25. Spectrograph of quasi-three-level laser at temperature of 60°C.

Table 4.8. Linewidth of quasi-three-level transitions at different temperature

Spectra, nm	Linewidth at different temperature, nm			
	-30°C	0°C	30°C	60°C
938	1.57	1.83	2.18	2.64
946	1.41	1.50	1.87	2.38

Table 4.9. Line shift variation of quasi-three-level lines at different temperature

Spectra (nm)	Wavelength shift			
	-30°C	0°C	30°C	60°C
938	938.48	938.46	938.48	938.44
946	946.14	946.14	946.08	946.12

Table 4.10. Intensity variation of quasi-three-level lines at different temperature

Spectra, nm	Intensity at different temperature, nm			
	-30°C	0°C	30°C	60°C
938	7.39	7.17	7.09	6.31
946	9.07	8.09	7.86	6.57

The number of lines in the spectra of quasi-three-level laser is remained the same, only that the intensity and the linewidth are obviously different. Tabel 4.8 shows as the temperature increases, the spectral linewidth of quasi-three-level lines at 938 and 946 nm increases. Furthermore, increasing the temperature causes thermal population in the ground sublevels. As a result spontaneous one phonon and muti-phonon emission processes effect on linewidth of inter stark transitions which terminate to the ground level.

In fact, the higher temperature causes the higher thermal vibration of ions in the crystals. Then varying in the crystal field owing to fluctuating inactive ions positions and their surrounding ligands bring about the local strain. When this strain is dynamically produced by the lattice vibrations, the interaction between the Nd^{3+} ions and the local crystalline field causes temperature-dependent broadening of the energy levels of the ions. This phenomenon is the reason of changing in linewidth of emission lines of the spectrum.

Beside the crystal field weaken by lattice thermal expansion, the phonon transition has its effect on the energy of electron-phonon system and modulation of Coulomb interaction between electrons and spin-orbital coupling. All of these have their contributions to the spectral thermal line shift.

Table 4.9 shows line positions are constant as temperature increases. In the temperature range of -30 to 60°C, systematic results were not obtained to investigate how line shift of Nd:YAG varying with temperature. Maybe this is because of a small range of temperature used in this present work.

Table 4.10 shows the intensity of 938.5 and 946.0 nm transition lines are temperature dependent. By increasing the temperature intensity of the 938.5 and 946.0 nm lines decrease monotonically. This is attributed to more re-absorption of these lines due to the thermal population at the ground level and therefore stimulated emission cross section of quasi-three-level emission lines are decreasing. Temperature dependence of stimulated emission cross section will be discussed in the next section in detail.

4.6. TEMPERATURE AND INPUT ENERGY DEPENDENCE OF THE 938.5 AND 946.0 NM STIMULATED EMISSION CROSS SECTION OF Nd^{3+}:YAG

Thermal broadening is caused by interaction between impurity ions with lattice vibrations in crystalline solids. At high frequencies, the ions are surrounded by the thermal vibrations of the lattice and then the atoms resonance frequency will modulate. This modulation can have several types of effects on the optical properties of the active ions. For instance, it can change the position of the electronic levels, thus leading to a broadening in the peak position of the spectral transitions.

Temperature and input energy dependency of linewidth causes a change of peak stimulated emission cross-section. However, to the best of our knowledge, few studies have been conducted on the lines at 938.5 and 946.0 nm wavelength induced by a flashlamp pumped Nd:YAG laser. Therefore, in this study experimental results of Nd:YAG crystal stimulated emission cross section at these wavelengths affected by the temperature and the input energy was investigated.

The recorded fluorescence spectra of quasi-three-level laser-induced from xenon flashlamp pumped Nd:YAG laser was depicted in Figure 4.26. Two strong peaks signals are clearly seen at wavelengths of 938.5 and 946.0 nm along with many weaker peaks. The peaks to produce blue light and their temperature and input energy dependence is interest to us. These two peaks are attributed to the transitions from Stark sublevels R_1 and R_2 in $^4F_{3/2}$ to Z_5 in $^4I_{9/2}$ intermanifold level.

The analysis of Figure 4.26 indicates that the $^4F_{3/2}$ manifold of Nd^{3+}:YAG is split into the Stark components R_1 and R_2 by 84.3 cm^{-1}, $\beta(R_1\text{-}Z_5) = 0.316$ and $\beta(R_2\text{-}Z_5) = 0.252$. The inter-stark radiative probability can be calculated using Eq. (2.16) substituting the values for intermanifold and inter-Stark branching ratios and radiative lifetime in addition to the values of Δ, k, and T. Finally the peak emission cross section for the inter-stark transitions at 938.5 and 946.0 nm are obtained via Eq. (2.17).

The cross-section of the two major lines from this quasi-three-level emission was estimated by measuring the linewidth at each temperature. The emission cross section at 938.5 and 946 nm were obtained at fixed electrical input energy keeping the voltage constant (500 V). However, the stimulated emission cross section over the temperature (-30 to +60°C) is measured and is shown in Figure 4.27. The emission cross section decreases monotonically as the temperature is increased from -30 to 60°C. The slopes were found to be -3.1×10^{-21} and -3.4×10^{-21} cm^2/°C for the 938.5-nm and 946-nm lines, respectively.

This decrement is because of the thermal broadening of the observed lines.

Finally, we turn our attention to the input energy dependence of fluorescence emission cross section. The emission cross section at a fixed temperature (20°C) at varying input energies 18–75 J is presented in Figure 4.28. The input energy is varied by manipulating the capacitor voltage of the flashlamp driver.

It can be seen from the figure the stimulated emission cross section for 938.5 and 946.0 nm decrease continuously as the input energy are increased that yield slopes of -4.4 × 10^{-21} and -6.8 × 10^{-21} cm^2/J for the two lines.

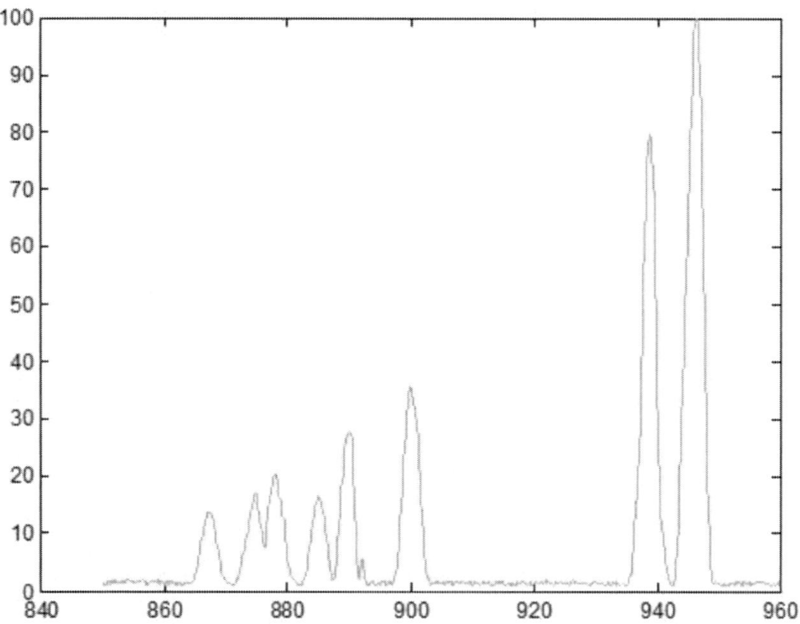

Figure 4.26. Normalized fluorescent emission spectra from $^4F_{3/2}$ to $^4I_{9/2}$ of Nd^{3+}:YAG.

Figure 4.27 and 4.28 indicate that the higher the temperature and pumping energy, the lower the production of photon flux at 938.5 and 946.0 nm (quasi-three-level) lines. This implies that thermal population at the ground level $^4I_{9/2}$ is increasing and as a result, much more transitions at 938.5 and 946.0 nm emissions will be absorbed and less photon at these wavelengths will be emitted as the fluorescence spectrum. In addition effects of temperature and input energy on emission cross section of 946.0 nm transition line is more dominant than on emission cross section on 938.5 nm transition line.

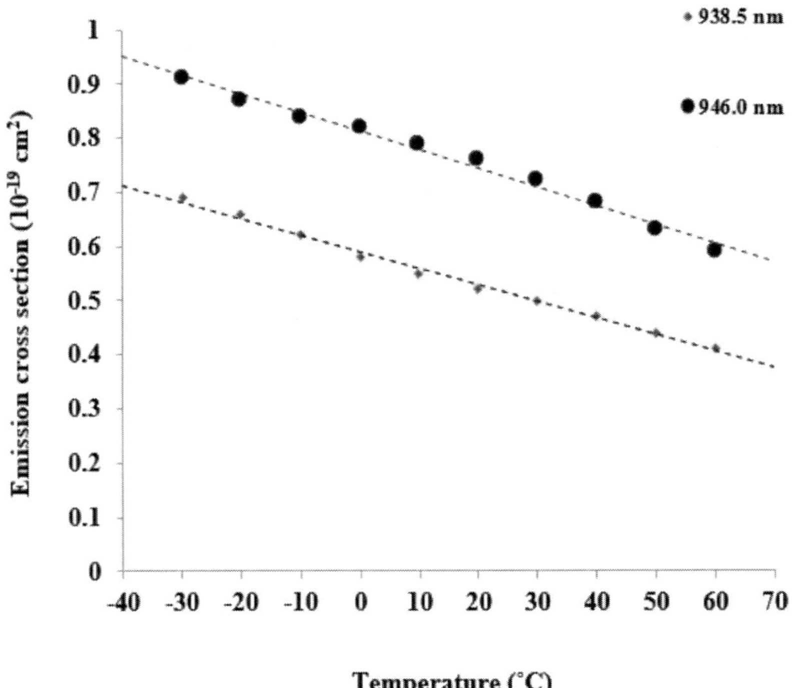

Figure 4.27. Peak stimulated emission cross section at 938.5 and 946.0 nm of Nd^{3+}:YAG versus temperature.

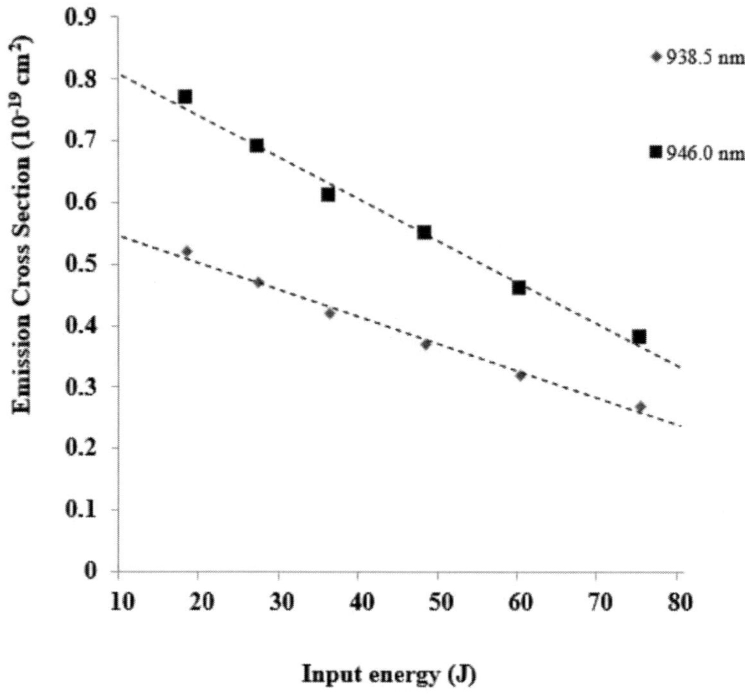

Figure 4.28. Input energy dependence of stimulated emission cross-section.

4.7. CHARACTERISTICS OF OPTICAL RESONATOR MIRRORS

The laser resonator was configured in the Plano-concave design in the present study. It is very difficult to obtain HR coating at 946 nm and HT coating at 1064 nm on pumping facet of the gain medium at the same time, but coatings of partial transmission at 946 nm and HT at 1064 nm can be obtained easily [1].

The emission cross section of the 946 nm transition line is much smaller than those at 1061, 1064 and 1319 nm. It was not likely to obtain laser action at 946 nm without suppression of feedback at these wavelengths. Thus, the resonator mirrors must have a small reflectivity at these wavelengths.

By separate installation of both high reflectivity and output, coupler mirrors the spectrum lines of Nd:YAG rod while they passed through the mirrors were recorded. By comparing the intensities of Nd:YAG emission lines before and after alignment of the mirrors characteristics of the rare mirror (high reflectivity) and output coupler (partial reflectivity) can be obtained.

Figure 4.29 shows the spectrum of Nd:YAG in voltage of 550 V before the alignment of the mirrors. The intensities of main transition lines which were detected by CCD camera and analyzed by a wavestar software were summarized and tabulated in Table 4.11.

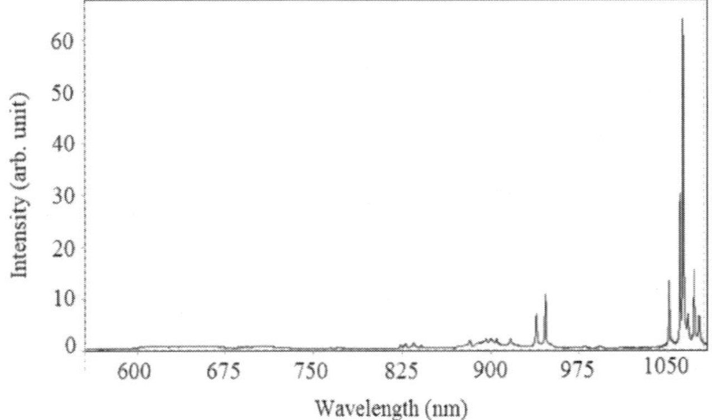

Figure 4.29. Spectrum of Nd:YAG transitions at a voltage of 500 V.

Table 4.11. Intensity of emission lines at voltage of 550 V

Wavelength (nm)	Intensity (arb. unit)
650.06	0.64
900.00	1.93
938.44	6.62
946.01	7.58
1051.83	13.03
1061.24	30.05
1063.92	63.67
1067.94	6.79
1073.46	15.24
1077.63	6.32

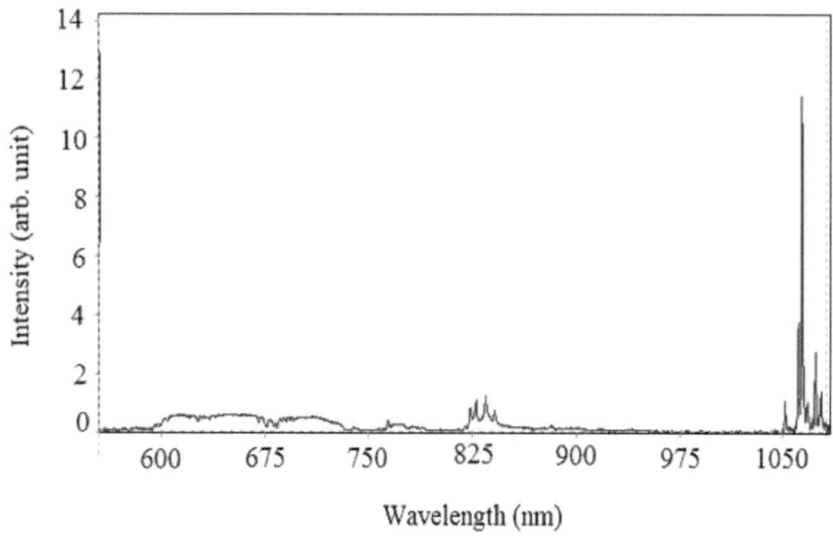

Figure 4.30. Spectrum of Nd:YAG emission after passing through HR mirror.

Table 4.12. Intensity and transmission of emission lines via HR mirror at 946 nm

Wavelength (nm)	Intensity (arb. unit)	Transmission (%)
650.06	0.55	85
900.00	0.11	6
938.44	0.03	0.5
946.01	0.00	0
1051.83	1.03	8
1061.24	3.72	12
1063.92	11.58	18
1067.94	0.97	14
1073.46	2.71	18
1077.63	1.38	22

Figure. 4.30 represents the spectrum of emission lines of Nd:YAG after they passed through the rare mirror. The intensities and transmissions of emission lines due to rare mirror were tabulated in table 4.12.

The optical reflectors were specially designed for operation at a wavelength of 946 nm. The HR reflector had an average transmission of more than 85% at 600–750 nm. It had an average transmission of around 15% at 1050–1080 nm. The reflectivity for 938 and 946 nm transition lines was 99.5% and nearly 100% respectively.

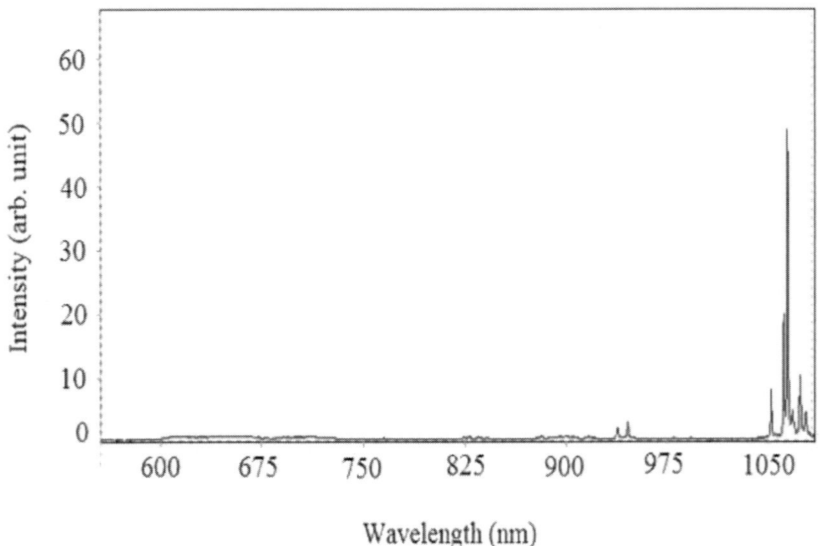

Figure 4.31. Spectrum of Nd:YAG emission after passing through 75% output mirror.

Table 4.13. Intensity and transmission of emission lines by 75% mirror coating at 946 nm

Wavelength (nm)	Intensity (arb. unit)	Transmission (%)
650.06	0.50	78
900.00	0.56	29
938.44	1.91	29
946.01	1.75	23
1051.83	7.94	61
1061.24	19.80	60
1063.92	48.73	77
1067.94	4.57	67
1073.46	10.21	67
1077.63	4.39	69

Figure. 4.31 shows the spectrum of emission lines of Nd:YAG after they passed through the mirror with a reflectivity of 75%. The intensities and transmissions of emission lines due to OC mirror were tabulated in Table 4.13.

The experimental results depict that the partial reflector (output coupler) had an average transmission of 80% at 600-750 nm and 65% at 1050–1080 nm, respectively. The output mirror had a measured reflectivity of only 23% at 1064 nm to suppress laser oscillation at this wavelength. In addition, this mirror had a reflectivity of nearly 70% and 75% for transition lines at 938 and 946 nm transition lines.

By using these mirrors, the problem of suppressing the much stronger lines especially 1064 nm parasitic oscillation could be solved. However, transition lines of 938.5 and 946.0 nm are near together and separation of them is difficult and required either an inter-cavity etalon or an outer-cavity prism. Therefore by using mentioned mirrors simultaneous generation of a dual wavelength at 938.5 and 946.0 nm was achieved.

4.8. QUASI THREE-LEVEL LASER TRANSITION PERFORMANCE

To satisfy the requirement of simultaneously multiple laser line oscillation, the special design of the laser oscillation cavity is necessary to control and to reduce the gain competition among the multiple wavelength lines of the gain medium.

The experimental setup after alignment the mirrors was depicted in Figure 3.11. By using the resonator mirrors introduced in the preceding section the condition of simultaneous oscillation of dual wavelengths at 938.5 and 946.0 nm laser lines were met by the equation of 2.85.

A typically recorded laser radiation spectrum of simultaneous generation of dual-wavelength oscillation at 938.5 and 946 nm by xenon flashlamp pumped Nd:YAG laser is depicted in Figure 4.32. The emissions lines have Lorentzian lineshape function which indicates the homogeneous broadening. These two peaks are attributed to the transitions from Stark sublevels R_1 and R_2 in $^4F_{3/2}$ to Z_5 in $^4I_{9/2}$ intermanifold level.

The dependency of the output of the long pulse Nd:YAG laser upon temperature is investigated at various input energies of the flashlamp used to pump the Nd^{3+}:YAG laser rod. This means besides verification of temperature the laser performance was also calibrated by measuring spectroscopically the laser output at different input energy. In practice the temperature of the cooling system is sustained at a constant value, meanwhile, the intensity of the spectrum is recorded for every increment of the input energy. The same procedure was repeated for next increment in temperature. The input energy is varied in the range of 18 – 75 J by manipulating the capacitor voltage of the flashlamp driver from 500 to 1000 V.

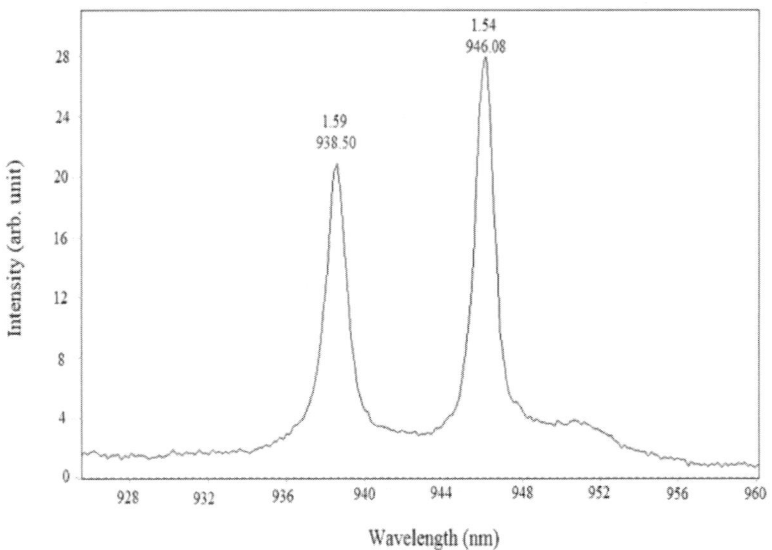

Figure 4.32. Spectrum of quasi-three-level laser radiations at 938.5 and 946.0 nm.

The quasi-three-level laser performances at various temperatures and energy are presented in Figures. 4.33 and 4.34. The laser performance at 938.5 nm (Figure 4.33) and 946.0 nm (Figure 4.34) are also found to be inversely proportional to temperature. These indicate that the higher the temperature, the lower the production of photon flux of quasi-three-level laser lines at 938.5 and 946.0 nm. This implies that the temperature of the cooling system did affect the thermal population at the ground level $^4I_{9/2}$. In another word, more re-absorption at the laser transitions lines at 938.5 and 946.0 nm emissions occurred. As a result, less photon at these wavelengths will be emitted as the fluorescence spectrum subsequently the laser radiation. Furthermore, the effects of temperature on emission cross section of 946.0 nm transition line are more dominant than on 938.5 nm transition line. This is based on drastic changes of emission cross section corresponding to change in the temperature.

Figures 4.35 and 4.36 reveal that the output (intensity of the spectrum) laser is increasing with the input energy. This implies that higher input energy produces more laser output.

Figures 4.35 and 4.36 also show the linear relationship between the output and input of long pulse for quasi-three-level Nd:YAG laser radiation. These experimental results agree well with theoretical of Eq. (4.58) which showed that the output laser is proportional to the input energy.

The higher the input energy produced higher laser output. The slope efficiency is unchanged with temperature. The average of slope efficiency for line 946.0 nm is almost constant as 11% and 10% for 938.5 nm. There are also notified that the laser performance becomes decreases as the temperature increases.

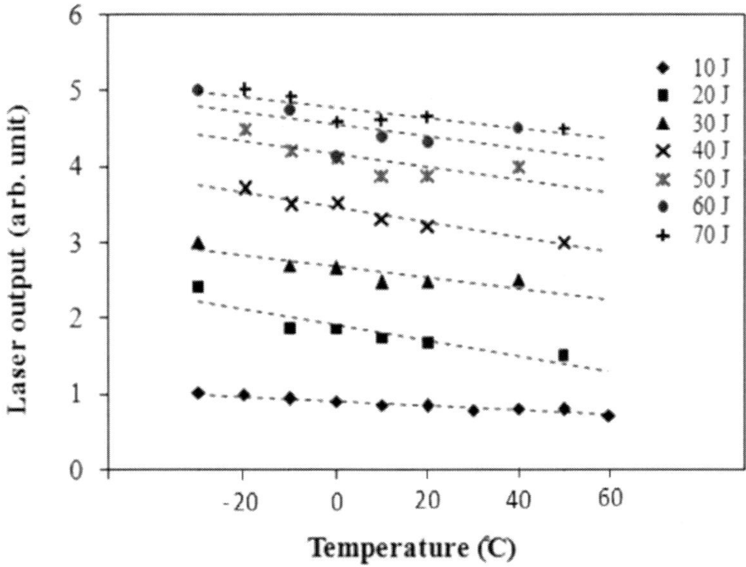

Figure 4.33. Output long pulse Nd:YAG laser (938.5 nm) temperature dependence.

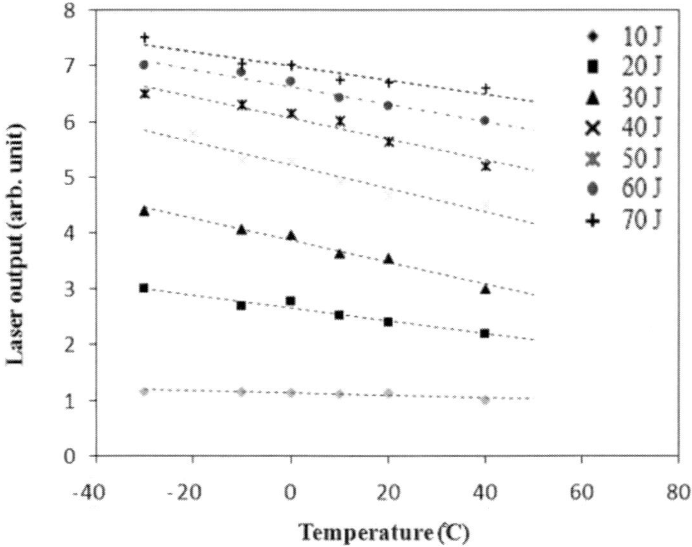

Figure 4.34. Output long pulse Nd:YAG laser (946 nm) temperature dependence.

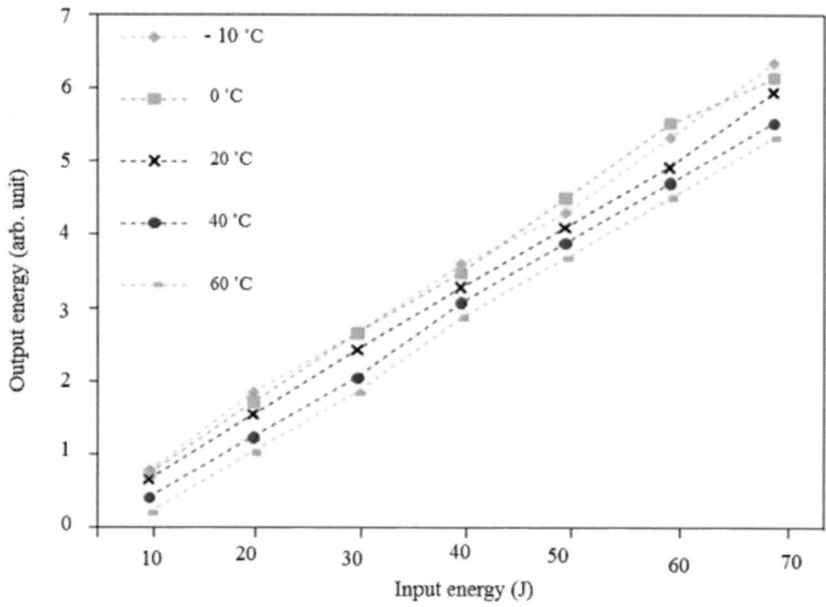

Figure 4.35. Long pulse Nd:YAG laser performance at 938.5 nm.

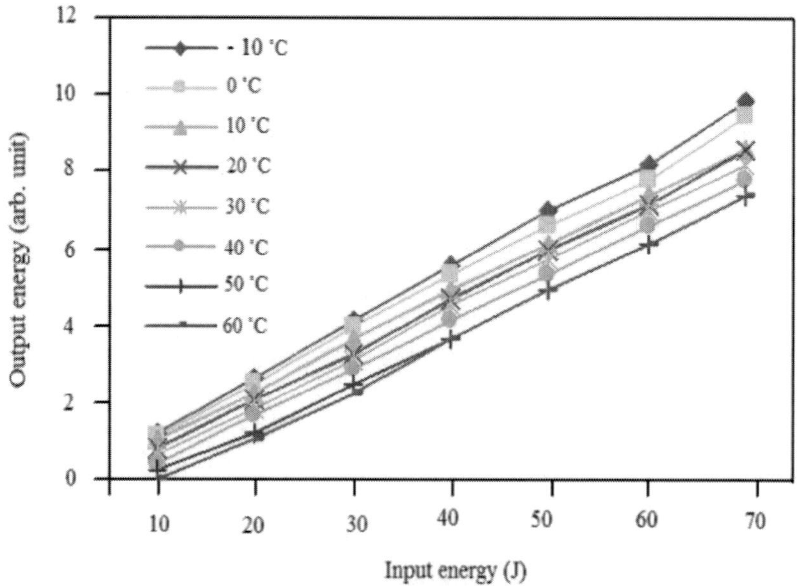

Figure 4.36. Long pulse Nd:YAG laser performance at 946 nm.

4.9. SUMMARY OF THE Nd:YAG LASER ROD PUMPED BY A NOVEL FLASHLAMP

The optical properties of Nd:YAG laser rod pumped by a novel flashlamp is studied by spectroscopic technique. The spectrum of fluorescence radiation indicates that the absorption line is a peak at 882 nm, implied that the neodymium ions is directly pumped by thermal boosted line. Consequently, the stimulated emission cross-section of 946 nm is found three times greater than the previously reported value at room temperature. Few studies have been conducted on the 938 and 946nm wavelengths induced by a flashlamp pumped Nd:YAG laser. Therefore, in this study, we reported the experimental evidence that stimulated emission cross section of Nd:YAG crystal at these wavelengths was affected by the temperature and the input energy. In addition, input energy dependency of intensity, linewidth, and wavelength position of quasi-three-level and four level system transitions was investigated. However, to the best of our knowledge similar studies have not been reported up to now.

Finally, in the present research simultaneous oscillation of dual-wavelength Nd:YAG laser at 938 and 946 nm pumped by flashlamp was introduced. Besides, laser performance of quasi-three-level transitions at 938 and 946 nm versus temperature and input energy was studied for the first time. The spectral line broadening and positions of the sharp emission lines for the three and four level inter-Stark transitions within the respective intermanifold transitions of $^{4}F_{3/2} \rightarrow {}^{4}I_{9/2}$ and $^{4}F_{3/2} \rightarrow {}^{4}I_{11/2}$ of Nd^{3+} ions in Nd:YAG crystal is investigated as a function of input energy. The linewidths of these transitions are increased with the input energy delivered on the target of Nd:YAG rod pumped by flashlamp. The emission lines for the three-level system of 938.5 nm and 946 nm transition lines shifted toward the longer wavelength (red shift), while the emission line for the four-level system transition lines remained unchanged by increasing the input energy. The experimental results of the input energy dependence of linewidths and line shifts of quasi-three-level transitions are explained in thermal population and one phonon and multi-phonon process in ground level. Input energy dependency of linewidth and line position of quasi-three-level and four level systems was reported in present work for the first time.

In this research, we reported the experimental evidence that stimulated emission cross section of Nd:YAG crystal at these wavelengths was affected by the temperature and the input energy. The stimulated emission cross-section for two lines of 938.5 and 946.0 nm are quantified as a function of temperature and input energy. The linear dependence of temperature and input energy is observed. The inter-stark emission shows Lorentzian line shape indicating homogeneous broadening. The thermal effect on laser transition at 946.0 nm is found more dominant in comparison to 938.5 nm.

Finally, simultaneous oscillation of dual-wavelength Nd:YAG laser at 938 and 946 nm pumped by flashlamp was introduced. Besides, laser performance of quasi-three-level transitions at 938.5 and 946.0 nm versus temperature and input energy was studied for the first time.

The quasi-three-level laser performances at various temperatures and energy are investigated. The laser performance at 938.5 nm and 946.0 nm are also found to be inversely proportional to temperature. These indicate that the higher the temperature, the lower the production of photon flux of quasi-three-level laser lines at 938.5 and 946.0 nm. This implies that the temperature of the cooling system did affect the thermal population at the ground level $^4I_{9/2}$. In another word, more reabsorption at the laser transitions lines at 938.5 and 946.0 nm emissions occurred. As a result, less photon at these wavelengths will be emitted as the fluorescence spectrum subsequently the laser radiation.

This is based on drastic changes of emission cross section corresponding to change in the temperature. In addition, the output (intensity of the spectrum) laser is increasing with the input energy.

This implied that higher input energy produced more laser output. The slope efficiency is unchanged with temperature. The slope efficiency for line 946 nm is almost constant as 10% and 9% for 938.5 nm. There are also notified that the laser performance becomes decreases as the temperature increases.

REFERENCE

[1]   Rui, Z., Zhiqiang, C., Wuqi, W., Xin, D., Peng, W. & Jianquan, Y. (2005). "High-power continuous-wave Nd:YAG laser at 946 nm and intracavity frequency-doubling with a compact three-element cavity", *Opt. Commu.*, *255*, 304-308.

Chapter 5

CONCLUSION AND FUTURE WORK OF ND:YAG LASER

ABSTRACT

A developed flashlamp power supply was utilized to produce the intense radiation of the Xenon flashlamp to pump the Nd:YAG laser rod. The purpose of this research was to investigate on the temperature dependence of quasi-three-level Nd:YAG laser performance. Spectroscopic properties of the Nd:YAG rod under pulsed flashlamp pumping was investigated from the output fluorescence spectrum of the flashlamp radiation and the Nd:YAG rod. These observations are new and may contribute towards the design architecture of lasers. The next issue is to improve the laser cavity and increase the input energy to recognize linewidth of all quasi-three-level lines. Then temperature and input energy dependence of these lines also can be investigated.

Keywords: spectroscopic results, Nd3+:YAG laser crystal, flashlamp

5.1. INVESTIGATION OF TEMPERATURE DEPENDENCE OF QUASI THREE LEVEL ND:YAG LASER PERFORMANCE

Nowadays most simultaneous generation of multi-wavelength is generated by diode pump solid state lasers. The disadvantage of these operations is that they are continuous or quasi-continuous. Thus they are difficult to control as well as they have low peak power.

A developed flashlamp power supply was utilized to produce the intense radiation of the Xenon flashlamp to pump the Nd:YAG laser rod. Xenon flashlamp radiation covers a wide range of electromagnetic spectrum from ultraviolet to near IR wavelength. Since only some wavelengths of light are absorbed, other wavelengths of light emerge as heat on the laser rod. This thermal effect can dramatically influence the laser performance.

For terrestrial applications, laser systems are mainly used in the temperature range from -60 to 60°C. The optical elements in a typical laser resonator (e.g., mirrors, beam splitters, etc.) show no variation of optical properties over a wide range of temperature. However, ambient temperature and the heat generated by the gain mediums of flashlamp pumping leads to thermal broadening and shift of laser lines which seriously affect on gain amplification, threshold power, frequency stability, and thermal tunability of the lasers and obstruct the laser operation. Furthermore, in most high power solid state lasers variation of emission cross section with temperature has a serious effect on the laser performance. As a result, extra heat will damage intracavity optics and varies stability of the output energy over the temperature range of interest.

Since Nd^{3+}:YAG laser performance is dependent on the temperature and pumping energy, careful investigation on the phenomena of linewidths and shifts of several lines of $^4F_{3/2}{\rightarrow}^4I_{11/2}$ and $^4F_{3/2}{\rightarrow}^4I_{11/2}$ intermanifold transition lines is required. In addition changes of laser output energy with temperature also is a crucial manner in solid-state laser materials. However, with accurate information about heat effects on active mediums, it would be possible to deal with the varying stability of the output energy over the temperature range of interest.

The purpose of this research was to investigate on the temperature dependence of quasi-three-level Nd:YAG laser performance. The system of our experiment includes three parts. The first part was energizing a laser crystal and stabilizing the stimulating emission of fluorescence radiation. The second part was measurement equipment comprises of the spectrometer. The third part was an alignment of cavity mirrors to achieve the laser output wavelengths.

A commercial laser rod Nd:YAG laser crystal is utilized as a gain medium. The doping level of the laser rod is 1 at.% with a dimension of 4 mm in diameter and 70 mm in length. The laser rod is enclosed in a ceramic reflector. The laser rod is placed parallel to a linear flashlamp filled xenon gas at 450 Torr. The flashlamp is pumped by a homemade power supply. The driver is triggered by simmer mode technique. A capacitor bank with the capacitance of 150 μF is charged by the maximum voltage of 1000 V thus the input energy is varied between 5 to 75 J.

The Nd:YAG crystal together with the flashlamp is flooded with a coolant comprised of the mixture of 60% ethylene glycol and 40% distilled water. Such particular coolant covers the range of temperature from -30°C to +60°C. A thermocouple is connected to the heat sink of alumina ceramic to measure the temperature of the cooling system.

The fluorescence radiation after pumping is emitted at one end of the laser rod. The light is detected by a CCD camera. The spectroscopic properties were analyzed via a Wavestar version 1.05 software. The resolution of this detection system is 0.5 nm so it can resolve most of the transition lines appeared from the pumping rod.

Spectroscopic properties of the Nd:YAG rod under pulsed flashlamp pumping was investigated from the output fluorescence spectrum of the flashlamp radiation and the Nd:YAG rod. The linewidth of each fluorescence lines was measured for estimation of effective emission cross section and saturation intensity.

In addition The influence of temperature and input energy on fluorescence emission cross section of Nd^{3+}:YAG crystal was studied. The cross-section was found to decrease as the temperature and the input energy was increased. The inter-stark emission shows Lorentzian line shape indicating homogeneous broadening. This was attributed to the thermal broadening mechanism of the emission line.

The spectral widths and shifts of the emission lines for the three and four level inter-Stark transitions within the respective intermanifold transitions of $^4F_{3/2}{\rightarrow}^4I_{9/2}$ and $^4F_{3/2}{\rightarrow}^4I_{11/2}$ were investigated over the range of 18 to 75 J. The emission lines for the $^4F_{3/2}{\rightarrow}^4I_{9/2}$ transitions shifted toward the longer wavelength (red shift) and broadened, while the positions and linewidths for the $^4F_{3/2}{\rightarrow}^4I_{11/2}$ transitions remained unchanged with increasing input energy. This was due to thermal population and one-phonon emission process in the ground state.

In order to achieve the dual wavelength laser operating at 938.5 and 946.0 nm, the coating for rare mirror had high reflectivity at 946.0 nm and high transmission more than 80% at 1064 nm and output coupler had a reflectivity of 75% at 946.0 nm and high transmission around 70% at 1064 nm. By using these mirrors, the problem of suppressing the much stronger lines especially 1061 and 1064 nm parasitic oscillations were solved.

Finally temperature dependence of quasi-three-level laser transitions for long pulse Nd:YAG laser was quantified. The output energy for two laser transitions of 938.5 and 946.0 nm was found to be inversely proportional to temperature and the slope efficiency was unchanged with temperature. It was also seen that the laser performance decreases as temperature increases.

These observations are new and may contribute towards the design architecture of lasers.

5.2. FUTURE WORK OF ND:YAG OPERATING LASER SYSTEMS

Our experimental work investigated the temperature dependence of some Nd:YAG laser parameters. However, some issues still remain to enhance this research.

By exposing the Nd:YAG crystal in the wide range of temperature, changes of widths and shifts of quasi three lines can be explained using Debye model in crystalline solids. Besides Debye temperature of quasi-three-level lines of 938 and 946 nm lines can be obtained.

The next issue is to improve the laser cavity and increase the input energy to recognize linewidth of all quasi-three-level lines. Then temperature and input energy dependence of these lines also can be investigated.

Another issue is to insert an intracavity etalon to achieve 938 nm transition line separately. In addition to using nonlinear crystals the blue light via frequency doubling of 938 and 946 nm, laser lines can be obtained.

ABOUT THE AUTHORS

Iraj S Amiri, PhD

Computational Optics Research Group, Advanced Institute of Materials
Science, Ton Duc Thang University, Ho Chi Minh City, Vietnam
Faculty of Applied Sciences, Ton Duc Thang University,
Ho Chi Minh City, Vietnam
Email: irajsadeghamiri@tdt.edu.vn

Dr. Iraj S Amiri, received his B. Sc (Applied Physics) from Public
University of Oroumiyeh, Iran in 2001 and a gold medalist M. Sc.
(Physics/Optics)) from University Technology Malaysia (UTM), in 2009.
He was awarded a PhD degree in Physics (photonics) in Jan 2014. He has
been doing research on several topics such as the optical soliton
communications, laser physics, plasmonics photonics devices, nonlinear
fiber optics, optoelectronics devices using 2D materials, waveguides,
quantum cryptography and nanotechnology engineering.

Dr. Abdolkarim Afroozeh

Lecturer
University of Larestan, Lar, Iran

Dr. Volker J Sorger

Department of Electrical and Computer Engineering
The George Washington University
Washington, D.C. 20052, USA

Dr. Xi Ling

Department of Chemistry
Division of Materials Science and Engineering
Boston University 590 Commonwealth Ave.
Boston, MA 02215, USA

Dr. Seyed Ebrahim Pourmand

Lecturer
Department of Optics and Lasers Engineering
Estahban Branch
Islamic Azad University
Estahban, Iran